Nanomaterials for Energy Applications

Topical Conference at the 2012 AIChE Annual Meeting

Pittsburgh, Pennsylvania, USA
28 October - 2 November 2012

ISBN: 978-1-62276-744-1

Printed from e-media with permission by:

Curran Associates, Inc.
57 Morehouse Lane
Red Hook, NY 12571

Some format issues inherent in the e-media version may also appear in this print version.

Printed by Curran Associates, Inc. (2013)

For permission requests, please contact AIChE
at the address below.

AIChE
3 Park Avenue
New York, NY 10016-5991

Phone: (203) 702-7660
Fax: (203) 775-5177

www.aiche.org

Additional copies of this publication are available from:

Curran Associates, Inc.
57 Morehouse Lane
Red Hook, NY 12571 USA
Phone: 845-758-0400
Fax: 845-758-2634
Email: curran@proceedings.com
Web: www.proceedings.com

TABLE OF CONTENTS

(34a) Inorganic Ligand-Stabilized CuInSe2 Nanocrystals Used in Photovoltaic Devices

C. Jackson Stolle, Matthew G. Panthani, Taylor B. Harvey, Vahid Akhavan and Brian A. Korgel, Department of Chemical Engineering, The University of Texas at Austin, Austin, TX

Nanocrystal solar inks show promise for developing high efficiency photovoltaic (PV) devices at low cost. Currently, the best nanocrystal-based PV devices require high temperature sintering, which limits substrate flexibility and increases cost. One of the greatest obstacles to higher device efficiencies without high temperature processing relates to the poor electrical transport and charge extraction from nanocrystal films. In particular, the organic capping ligands used during synthesis to stabilize nanocrystals in solution create an electrically insulating barrier which hinders particle to particle charge transport and limits PV device performance. Replacing the organic molecules with inorganic capping ligands could greatly improve charge transport through nanocrystal films in PVs as well as device performance.

We report the use of inorganic metal chalcogenide-hydrazinium complexes (MCCs) as well as S^{2-}, HS^{-}, and OH^{-} ions as capping ligands in $CuInSe_2$ (CIS) nanocrystal-based photovoltaic devices. A ligand exchange is used to replace the organic capping ligand, leaving the MCC ligand-capped nanocrystals dispersed in hydrazine and the S^{2-} ion-capped nanocrystals dispersed in water. PV devices made from MCC ligand-capped CIS nanocrystals exhibited power conversion efficiency comparable to oleylamine-capped CIS nanocrystal PVs. S^{2-} ion-capped CIS devices exhibit lower efficiency, but show that nanocrystal PVs with inorganic ligands can be deposited using aqueous dispersions.

(34b) Molecular Simulations of Ligand-Capped Nanocrystals

Ananth P. Kaushik and Paulette Clancy, Chemical and Biomolecular Engineering, Cornell University, Ithaca, NY

Nanocrystalline solids have become the subject of intense study due to their unique optical properties and their capacity to form self-assembled superlattices. These properties make them suitable for use in a variety of applications such as solar cells, light-emitting devices, transistors, etc. The self-assembly of these nanoparticles is governed by interactions at the molecular level, and hence, understanding the nature of these interactions could be instrumental in achieving precise control over the self-assembly process. The nanocrystals are "capped" by organic molecules which are believed to drive the self-assembly and stabilize the final superlattice and which are responsible for most of the interparticle forces. Despite the importance of these ligand-ligand interactions, there is very little fundamental understanding of their role in structure-direction. We have used atomically and molecularly explicit Molecular Dynamics simulations to create this understanding. Specifically, we study the links between processing conditions and the resultant symmetry of the superlattice, whether face-centered cubic (fcc) or body-centered cubic (bcc) structures, in a close collaboration with the Hanrath group.

These results form the first report of computer simulations of experimentally relevant sized ligand-capped nanocrystals (3-6 nm) in contrast to prior simulation work in the literature (sub-3 nm). Our results for the free energies of the system provide definitive proof of a clear dependence of preferred superlattice symmetry on the ratio of ligand length to nanocrystal size. Importantly, we have also uncovered the two key molecular mechanisms that are at play to direct the eventual preference. These simulations have also shown that nanoparticle interactions are critically dependent on parameters such as ligand grafting density and ligand coverage on specific facets of the nanocrystals, which ultimately help govern the self-assembly process.

In our most recent work, we are investigating the effect of shape and size of nanocrystals on the electronic energy states, which are critically important to their performance as photovoltaic solar cells. Our preliminary results confirm experimentalists' suspicions that surface ligands play a role in modifying the electronic properties. We are investigating the properties of excitons by modeling excited states using Time Dependent Density Functional Theory (TD-DFT). We are also studying the electronic properties of nanocrystals coupled through organic linker molecules that will help inform the feasibility of this approach for experimentalists.

The impact of this kind of study is the provision of key insights into molecular-scale information about the relative roles of surface-bound ligand and nanoparticle cores that are impossible to determine experimentally. In addition, this insight can be leveraged in the future into more coarse-grained mesoscale simulations of the whole self-assembly process, namely nucleation and growth phenomena, which may then to be used to guide future experimental studies.

(34c) Nanocrystal Building Blocks in Photofunctional Superlattice Structures

Joshua Choi, Cornell University, Ithaca, NY, Kaifu Bian, Cornell University, Ithaca, Detlef Smilgies, Cornell High Energy Synchrotron Source, Ithaca, NY and **Tobias Hanrath**, Chemical and Biomolecular Engineering, Cornell University, Ithaca, NY

Recent advances in synthesis, characterization, and the emerging understanding of their size-dependent properties have created exciting opportunities for semiconductor nanomaterials to contribute to the development of next-generation energy conversion technologies. Semiconductor nanocrystal quantum dots are particularly attractive material candidates for the efficient capture of solar emission in inexpensive, thin film photovoltaic devices due to their large absorption cross sections, low-cost solution-phase processing and size-tunable energy gaps. Despite the immense promise, the development of cost- and performance-competitive nanocrystal-based solar cells has, to date, fallen short of expectations. This gap is primarily due to the current lack of answers to the challenge of how to efficiently extract photogenerated charges from quantum confined systems and transport them to external electrodes. In addition to the thermodynamic challenge of integrating the nanocrystal into a structure in which separation of photogenerated electron and hole can be separated and transported, there are also important kinetic challenges. Specifically, the rates of the sub-processes of separation and transport across the various interfaces must surpass competing recombination and trapping dynamics. We review our recent work investigating the exciton dissociation and charge transport between quantum confined nanocrystals as a function of the interparticle spacing. Beyond controlling the interdot spacing, directing the nanocrystals into ordered superstructures with predefined translational and orientational order is a critical parameter for controlling the electronic coupling in three dimensional superlattices. We will present recent experiments relating superlattice symmetry and optical properties of the ensemble.

(34d) Combinatorial Chemical Bath Deposition of $Cd_{1-x}Zn_xS$ for Thin Film Photovoltaics

Kevin M. McPeak, ETH- Zurich, Zurich, Switzerland and Jason B. Baxter, Department of Chemical and Biological Engineering, Drexel University, Philadelphia, PA

$Cd_{1-x}Zn_xS$ (CdZnS) is a promising replacement for the CdS buffer layers in copper indium gallium (di)selenide (CIGS) solar cells because the wider band gap of CdZnS offers improved optical transmittance of blue light. Chemical bath deposition (CBD) is the state-of-the-art deposition method for CdS and CdZnS. However, CBD is poorly understood, and relationships between bath composition and stoichiometry, microstructure, and optoelectronic properties of the deposit are lacking.

We introduce combinatorial chemical bath deposition (CBD), using a continuous flow microreactor, as a new technique to deposit ternary semiconductor thin films with graded composition along the length of a single substrate. The graded composition results in graded material properties; and spatially-dependent characterization of the substrate enables rapid and unprecedented understanding of the effect of growth conditions on optical, electronic, and structural properties of the material. The microreactor used behaves like a plug flow reactor whereby the bath composition changes as material is deposited on the substrate.

CdZnS films showed increasing Zn:Cd ratio, quantum confinement, and resultant monotonic increase in band gap from 2.42 to 2.75 eV over the length of a single substrate. X-ray diffraction and x-ray absorption spectroscopy indicate that the deposit is a single-phase solid solution, rather than phase-separated CdS and ZnS. Combinatorial CBD enables rapid identification of CdZnS compositions that are ideal candidates for thin film photovoltaics, as well as determination of the CBD conditions required to deposit them.

(34e) Improved Performance of Earth-Abundant Cu2ZnSn(SxSe1-x)4 Solar Cells Through Ge Incorporation

Charles Hages, Chemical Engineering, Purdue University, West Lafayette, IN and Rakesh Agrawal, School of Chemical Engineering, Purdue University, West Lafayette, IN

Improved Performance of Earth-Abundant $Cu_2ZnSn(S_xSe_{1-x})_4$ Solar Cells through Ge Incorporation

Charles Hages and Rakesh Agrawal

School Of Chemical Engineering, Purdue University, West Lafayette, IN, USA

Advancements in thin film $Cu_2ZnSn(S_xSe_{1-x})_4$ (CZTSSe) solar cells have recently achieved power conversion efficiencies of >10%, indicating the potential of this low cost, earth abundant material system as a viable alternative to copper indium gallium diselenide (CIGSe) and CdTe absorbers[1]. The initial appeal of CZTSSe as a viable solar p-type material comes from its ideal band gap of 1.0 to 1.5 eV. Band gap tuning in current high efficiency devices in literature has been optimized mainly through control of the S/(S+Se) ratio in CZTSSe. However, current research has recently shown the additional ability for tuning band gap of CZTSSe through partial substitution of Sn for Ge into the crystal lattice forming $Cu_2Zn(Sn_yGe_{1-y})(S_xSe_{1-x})_4$ (CZTGeSSe)[2,3]. This presentation will introduce material and electrical characterization of both CZTGeS synthesized nanoparticles as well as CZTGeSSe sintered films as a function of Ge incorporation.

Powder x-ray diffraction (PXRD) has shown that control of the reaction synthesis procedure allows synthesized CZTGeS nanoparticles to maintain tetragonal structure (kesterite/stannite) throughout all ratios of Ge incorporation. Current work suggests the importance of the tetragonal (kesterite) crystal structure for enhanced photovoltaic device performance, including CZGeS where the wurtzite phase has been calculated to have increased stability. Raman spectroscopy has also been measured for CZTGeSSe to understand changes in the main Raman shift mode of CZTSSe as Ge is incorporated into the crystal lattice. In addition, Raman spectroscopy has been used to help understand and identify the formation of binary and ternary phases in CZTGeSSe nanoparticles and sintered films.

Electrical characterization of CZTGeSSe devices has shown a linear increase in open-circuit voltage (Voc) as the Ge content in increased. The current optimization at 30% Ge/(Sn+Ge) has been achieved through a balance of increased Voc while maintaining a sufficiently high photo-generated current (Jph) as the band gap in increased with Ge content. Further analysis is under way using photoluminescence to quantify changes in optical absorption and band gap as a function of Ge incorporation.

Further analysis of temperature dependent Voc has shown the CdS/CZTGeSSe heterojunction interface as a source of limiting recombination in the measured devices. In addition to this, elemental losses of Ge at the interface have been found to significantly contribute to measured Voc limitations in CZTGeSSe devices, further suggesting the importance of the buffer/absorber interface for improvements in device performance. Changes in the reaction synthesis procedure as well device fabrication techniques have proven the ability for decreased elemental losses at the interface, with direct positive improvements seen in Voc measurements.

In addition to band gap optimization, Ge incorporation into CZTSSe has been used to minimize the unfavorable conduction band offset between CdS/CZTSSe. External quantum efficiency measurements have shown increased carrier collection efficiency due to a decrease in the conduction band offset as Ge incorporation is increased to 30%. This work demonstrates the importance of both S/Se control as well as Ge/Sn control to optimize the band gap as well achieve favorable band alignments.

[1] D. A. R. Barkhouse et al., "Device characteristics of a 10.1% hydrazine-processed Cu2ZnSn(Se,S)4 solar cell", *Prog. Photovolt: Res. Appl.*, **20**, 2012, pp. 6-11.

[2] G. M. Ford et al., "Earth Abundant Element Cu2Zn(Sn1-xGe)S4 Nanocrystals for Tunable Band Gap Solar Cells: 6.8% Efficient Device Fabrication", *Chem. Mater.*, **23**, 2011, pp. 2626-2629. 3.

[3] Q. Guo et al., "A generalized and robust method for efficient thin film photovoltaic devices from multinary sulfide nanocrysatal inks", *Photovoltaic Specialists Conference (PVSC), 2011 37th IEEE.* pp. 003522-003526.

(34f) Charge Separation in Extremely Thin Absorber Solar Cells

Hasti Majidi, Glenn W. Guglietta II, Leah C. Spangler and **Jason B. Baxter**, Department of Chemical and Biological Engineering, Drexel University, Philadelphia, PA

Extremely thin absorber (ETA) solar cells utilize a very thin light-absorbing layer at the nanostructured interface between n- and p-type semiconductors to enable efficient separation of photoexcited charge carriers. The high-area interface allows smaller absorber thicknesses to harvest the same amount of light as thicker planar cells. Because the absorber is thin, efficient charge separation can be achieved with material of lower electronic quality, which enables low-cost processing and new materials not possible with planar cells. Here we report on our recent efforts to understand and control interfacial electron and hole transfer in ETA cells consisting of ZnO/CdSe/CuSCN and ZnO/CdSe/electrolyte, in both nanowire-based and planar configurations. Nanowire arrays provide large surface area and conduits for 1-D electron transport, while planar cells provide simplified geometry for model studies. Our approach utilizes a combination of solar cell measurements and ultrafast transient absorption spectroscopy to understand the effects of CdSe thickness, annealing conditions, and interfacial treatments on the dynamics and efficiency of charge carrier separation, and ultimately on the solar-to-electric energy conversion efficiency. These studies provide guidelines for architecture design and materials selection for ETA solar cells.

(103a) Graphene Aerogel-Based Counter Electrodes for Dye- Sensitized Solar Cells

Wei-Yung Cheng[1], Chun-Chieh Wang[2] and Shih-Yuan Lu[1], (1)Dept. of Chemical Engineering, National Tsing-Hua University, Hsin-Chu, Taiwan, (2)Dept. of Chemical Engineering, National Tsing-Hua University, Hsinchu, Taiwan

Graphene aerogels were prepared following an organic sol-gel process developed by Worsley et al. [Ref. 1] with some minor modifications. Under preferred reaction conditions, the graphene aerogels obtained were with a high specific surface area of 814 m^2/g and a high electric conductivity of 850 S/m, and were applied as a counter electrode material for dye-sensitized solar cells. The performance of the graphene aerogel in dye sensitized solar cells was found to be dependent on its film thickness, with thicker films offering more surface areas for the involved catalytic reaction but at the same time increasing the charge transport resistance. At a graphene aerogel film thickness of 4.9 μm, a power conversion efficiency of 96% of that achieved with the Pt counter electrode based DSSC was obtained. In addition, a thinner graphene aerogel film of 1.7 μm, when loaded with Pt of 1 mole% through a light reduction procedure, achieved a power conversion efficiency of 98% of that obtained with the Pt counter electrode based DSSC. The excellent performances of the graphene aerogel-based counter electrodes were manifested with electrochemical impedance analyses and cyclic voltammetry catalytic activity analyses.

Ref. 1: M.A. Worsley, P.J. Pauzauskie, T.Y. Olson, J. Biener, J.H. Satcher, Jr., T.F. Baumann, J. Am. Chem. Soc., **132**, 14067 (2010).

(103b) Graphene Metal Oxide Nanoassemblies for Dye-Sensitized Solar Cells

Paul A. Charpentier[1], Nasrin Farhangi[2] and Serge Ayissi[2], (1)Department of Chemical and Biochemical Engineering, Western University, London, ON, Canada, (2)Chemical & Biochemical Engineering, Western University, London, ON, Canada

Tremendous interest exists towards synthesizing nanoassemblies for dye-sensitized solar cells (DSSCs) using earth abundant and friendly materials with green synthetic approaches. Recently, two-dimensional graphene has initiated a rapid exploration of its potential in a broad range of fields including energy conversion and catalysis [1]. Catalyst particles can be incorporated directly onto an individual graphene sheet, providing enhanced properties for DSSCs [2]. Graphene is a good electron acceptor and can serve as charge trapping sites and reduce the electron-hole recombination rate to enhance the photoefficiency of titania[3].

In this research, uniform TiO_2 nanowires were grown on the surface of graphene sheets using a sol-gel method in supercritical carbon dioxide ($scCO_2$) to help exfoliate graphene while preventing agglomeration and controlling the particle size. Fe doped TiO_2 on the graphene sheets were also prepared in both ethanol and $scCO_2$ to extend its band gap to the visible region.[4] TiO_2 nanoparticles less than 5 nm (when ethanol was used as a solvent) and TiO_2 nanowires less than 40 nm (when $scCO_2$ was used as a solvent) were uniformly decorated on the graphene sheets. Both doped TiO_2 nanoparticles and nanowires showed smaller crystal size, and higher visible absorption, surface area, and higher efficiency solar cells compared to similar materials without graphene.[5] Theoretical band structures have been studied using the Vienna ab-initio Simulation Package (VASP)[6] based on the Density Functional Theory (DFT). Band gap and adsorption energy values of structural TiO_2 were calculated respectively in the bulk and on graphene by physisorption and chemisorption. Functionalized graphene sheets decrease significantly the band gap but increase the binding energy values, due to carboxylate adsorption sites, of structural TiO_2 compounds which facilitate DSSC performance.

[1] I.V. Lightcap, T.H. Kosel, P.V. Kamat, Nano Letters 10 (2010) 577-583.

[2] H. Zhang, X. Lv, Y. Li, Y. Wang, J. Li, ACS Nano 4 (2009) 380-386.

[3] X.Y.L.H.-P.C. Zhang, X-L; Linb, Y., Journal of Materials Chemistry 20 (2010) 2801-2806.

[4] Nasrin Farhangi, Yaocihuatl Medina-Gonzalez , Rajib Roy Chowdhury, and Paul A. Charpentier, *Nanotechnology* (2012), In Press.

[5] Nasrin Farhangi, Rajib Roy Chowdhury, Yaocihuatl Medina-Gonzalez, Madhumita B. Ray and Paul A. Charpentier, Applied Catalysis B: Environmental 110 (2011) 25– 32.
[6] G. Kresse, D. Joubert, Phys. Rev. B 59 (1999) 1758

(103c) Dye-Sensitized Solar Cells Based On Vertically Ordered 1-D Nanostructures

Di Gao, Chemical and Petroleum Engineering, University of Pittsburgh, Pittsburgh, PA

Dye-sensitized solar cell (DSC) is one of the most promising devices for low-cost solar-to-electricity energy conversion. Although an energy conversion efficiency of 11% has been demonstrated for such cells, researchers are still seeking to further increase the efficiency by using alternative sensitizers and redox electrolytes, and to fabricate solid state or non-volatile liquid DSCs. The progress of these efforts, however, is impeded by the disordered structure of mesoporous TiO_2 films used in current DSCs. The mesoporous TiO_2 films, composed of sintered TiO_2 nanoparticles, induce a short electron diffusion length, which limits further improvement in DSC performance. The disordered porous structure also makes infiltration of solid or viscous electrolytes into the electrodes difficult and ineffective. Synthesis and fabrication of vertically aligned 1-D nanostructure directly on transparent conducting oxides (TCO) has been identified as a promising means to overcome these issues. Here we present recent progresses we have made toward high-performance DSCs based on vertically ordered 1-D nanostructures, including (i) preferential growth of long ZnO nanowire arrays on TCO, (ii) synthesis of TiO_2 nanotubes, (iii) synthesis of multilayer assembly of ZnO-TiO_2 core-shell nanostructures, and (iv) high-efficiency solid-state DSCs based on multilayer assembly of hybrid ZnO-TiO_2 nanostructures.

(103d) Multiscale Anode Assemblies for Improving Efficiency and Versatility of Dye-Sensitized Solar Cells

Pisist Kumnorkaew[1], Midhun Joy[2], James F. Gilchrist[1] and Mark A. Snyder[1], (1)Department of Chemical Engineering, Lehigh University, Bethlehem, PA, (2)Dept. of Chemical Engineering, Lehigh University, Bethlehem, PA

The operation and efficiency of dye-sensitized solar cells (DSSCs) are inextricably linked to both light and charge transport/management. Only incremental improvements in device efficiency have been realized since the pioneering work by Grätzel [1] more than two decades ago despite significant optimization of sensitizer efficiencies. Improvements in overall device efficiencies tend to be confounded by complexities associated with the multicomponent nature of DSSCs, and demand careful materials engineering to enhance both light and charge transport across these devices. In this talk we will describe efforts aimed at engineering anode micro/nanostructure as well as active area as means for improving device efficiencies. We will show how fundamental insight into efficiency enhancements associated with anode-incorporated microlens technologies and confinement of electron transport can be leveraged through assembly of multiscale anode platforms of controlled nano/microscale structure and patterned meso/macroscale ordering. The scalable manufacture of these multiscale anodes employs techniques spanning convective to electrophoretic assembly. In combination with device packaging strategies, we will show how these multiscale structured anodes can translate to improvements in overall cell efficiency, and how progress towards realization of more versatile flexible devices can be realized.

[1] O'Regan, Grätzel, *Nature*, 353 (1991).

(103e) High-Efficiency Solid-State Dye-Sensitized Solar Cells Based On TiO_2-Coated ZnO Nanowire Arrays

Chengkun Xu, Jiamin Wu, **Umang V. Desai** and Di Gao, Chemical and Petroleum Engineering, University of Pittsburgh, Pittsburgh, PA

With increasing concern over the twin crises of fossil fuel depletion and environmental degradation, dye-sensitized solar cells (DSCs) have received considerable interest as one of the most promising devices for converting solar energy to electricity at low cost. Replacing the liquid electrolytes in DSCs with solid-state hole-transporting materials (HTMs) may solve the packaging challenge and improve the long-term stability of DSCs. The efficiencies of such solid-state DSCs (ss-DSCs), however, have been far below the efficiencies of their counterparts that use liquid electrolytes, primarily due to the challenges in filling HTMs into thick enough sensitized films based on sintered TiO_2nanoparticles. One approach to solving the challenge of filling sensitized films with solid HTMs is to replace the sintered nanoparticle-based mesoporous TiO_2 film with vertically ordered nanostructures. Here we report fabrication of high-efficiency ss-DSCs using multilayer TiO_2-coated ZnO nanowire arrays as the photoanodes. The straight channel between the vertically aligned nanostructures combined with a newly developed multistep HTM filling process allows us to effectively fill sensitized films as thick as 50 μm with the HTMs. The resulting ss-DSCs yield an average power conversion efficiency of 5.65%.

(103f) Modeling of Solid and Particulate Passivating Layers in Dye Sensitized Solar Cells

Rocco Panella[1], B. Erik Ydstie[1] and Dennis C. Prieve[2], (1)Chemical Engineering, Carnegie Mellon University, Pittsburgh, PA, (2)Dept. of Chemical Engineering, Carnegie Mellon University, Pittsburgh, PA

Dye-sensitized solar cells (DSCs) utilize a porous layer of wide-bandgap semiconductor nanoparticles which have been deposited on a transparent conductive oxide (TCO) surface such as tin-doped indium oxide or fluorine-doped tin oxide. This porous layer, typically about 10-20 microns thick, is coated with a dye which is highly absorbent over the visible spectrum. This dye functions as the means of charge separation in these cells, injecting excited electrons into the porous TiO_2 layer and interacting with a redox couple suspended in a liquid electrolyte.

A common source of current loss in DSCs is the transfer of electrons from the TCO substrate to the liquid redox couple, occurring at the undesirable interface of the liquid and the TCO. This problem is typically addressed by depositing a passivating layer of TiO_2 on the bare TCO before the deposition of the porous layer. While this effective method for stopping TCO-electrolyte recombination is standard practice when building DSCs, there is little consensus on the specific impact of this passivating layer on DSC performance.

Our lab group has fabricated DSCs using passivating layers consisting of both solid-phase TiO_2 and particulate TiO_2, and we have compared these to results from other published works. We note that, when using passivating layers consisting of particulate TiO_2 which have been deposited via electrostatics, we observe a 9% improvement in open-circuit voltage and up to a 50% improvement in short-circuit currents as opposed to cells which are not passivated. Additionally, cells which are passivated using particulate layers show higher short-circuit currents than cells passivated using solid TiO_2, but have lower fill factors. We have used these data as an aid to the building of a theoretical and computational model for the steady-state and time-dependent operation of a DSC, with an emphasis on the effects of interactions occurring at the TCO-electrolyte interface. Our calculations have shown that by simply allowing for TCO-electrolyte interaction via a Butler-Volmer mechanism, we are able to account for commonly-cited, non-ideal current-voltage behavior of the anode in DSCs. Additionally, taking this interaction into account when modeling steady-state cell behavior allows for higher accuracy when calculating overpotential losses in the liquid components of the DSC. Time-dependent studies, solved numerically, have proven to be qualitatively similar to experimental impedance measurements, indicating that our additions to the modeling of DSCs carry physical significance. We believe that the differences in resistance, as well as the effective passivated area, between solid and particulate blocking layers results in the differences in their effect on DSC performance.

This work is the first suggestion for an analytical mechanism by which TCO-electrolyte interaction affects DSC performance, as well as an explanation for non-ideal anode behavior. While experimental optimization of DSC blocking layers is common practice, we hope that this work will begin more fundamentals-driven optimization.

(153a) Polymer-Free near-Infrared Photovoltaic with Single Chirality (6,5) Semiconducting Carbon Nanotube Active Layer

Rishabh Jain[1], Rachel Howden[2], Kevin Tvrdy[2], Steven Shimizu[2], Andrew J. Hilmer[2], Tom McNicholas[2], Karen K. Gleason[2] and Michael S. Strano[2], (1)Department of Materials Science and Engineering, Massachusetts Institute of Technology, Cambridge, MA, (2)Department of Chemical Engineering, Massachusetts Institute of Technology, Cambridge, MA

The incorporation of single-walled carbon nanotubes (SWNTs) into next generation solar cells as near infrared absorbers has demonstrated the potential to efficiently harness energy in the 1000nm to 1400nm range. However, SWNT-enabled photovoltaics (PVs) to date have required the use of polymers, which served as either SWNT-wrapping (isolating) agents or as direct components of the photoactive layer. While such layers are expected to increase device performance at the laboratory scale, the use of PV engineered polymers is oftentimes restrictive due to low photostability and the necessity of highly controlled environments for assembly and characterization. In contrast, carbon nanotubes have the strong advantage of being extremely stable in air while at the same time absorbing in the near infrared region of the solar spectrum. Here, for the first time, we demonstrate a polymer-free carbon based photovoltaic which relies on exciton dissociation at the SWNT/C60 interface. Devices yielded 0.5% external quantum efficiency (EQE) at the peak of the nanotube absorption spectrum and a power conversion efficiency of 0.10% under AM1.5 illumination. Interestingly, this value exceeds many of those reported for polymer/SWNT/PCBM composite based photovoltaics. Further, the addition of only 20% (6,4) SWNTs to an otherwise pure (6,5) SWNT film reduces device efficiency 30 times, demonstrating the importance of an energetically homogeneous film with respect to SWNT chirality. Through the construction of a carbon based solar cell completely free of polymeric active or transport layers, we show both the feasibility of this novel device as well as inform the mechanisms for inefficiencies in SWNT and carbon based solar cells.

(153b) Synthesis Routes of Phosphorous-Free Semiconducting Selenide Nanoparticles for Photovoltaic Applications

Bryce C. Walker, Steven McLeod, Bethlehem Negash and Rakesh Agrawal, School of Chemical Engineering, Purdue University, West Lafayette, IN

Selenium based semiconductors have filled large roles in photovoltaics with materials that include Copper-Indium-Gallium-Selenide (CIGSe), Cadmium Selenide, Copper Selenide, and Copper-Zinc-Tin-Selenide (CZTSe). These materials, and their other chalcogenide derivatives, are major players in thin-film photovoltaics for lower cost solar cells. Nanomaterials of these devices have been fabricated, but traditionally rely on either toxic selenium sources, or elemental selenium dissolved with phosphorous or boron containing chemicals. The hazards of toxic selenium sources should be avoided for use in scalable solar cell manufacturing. Meanwhile the addition of phosphorous and boron should be avoided in order to keep these components from contaminating the final device fabricated from these nanocrystals.

We have developed a method to enable the fabrication of selenide nanocrystals from elemental selenium without using any phosphorous or boron in the process. This process has allowed the formation of selenide nanocrystals in organic media with great size control, and for applications in scalable roll-to-roll device fabrication. By adjusting the S to Se ratio, band-gap control of the individual nanocrystals is also achieved. These nanocrystals are currently being investigated for their photovoltaic applications.

(153c) Tuning the Structure of Titania Films with Orthogonally Aligned Cylindrical Nanopores for Bulk Heterojunction Inorganic-Organic Solar Cells

Suraj Nagpure and Stephen E. Rankin, Chemical and Materials Engineering, University of Kentucky, Lexington, KY

The objective of this work is to synthesize 2D hexagonal close packed (HCP) titania thin films with 10 nm orthogonally aligned cylindrical nanopores for photovoltaic materials. HCP structures cast onto hydrophilic oxide surfaces (such as those of titania or indium tin oxide that might be present in bulk heterojuction materials) are usually found to align parallel to the surface. To provide vertical alignment of the pores, we use surface modification of substrates with crosslinked Pluronic PEO-(b)-PPO-(b)-PEO triblock copolymer surfactant F127, which also is used as the pore template to provide a large-pore HCP structure. Here we show that this approach leads to vertically aligned HCP structures in titania with relatively large nanopores, as determined by XRD, SEM and TEM characterization. For the binary system of F127 and water, a hexagonal lyotropic liquid crystal phase is obtained at weight ratio of F127: water of 1.7 to 3. By replacing the volume of water with an equivalent volume of hydrolyzed titanium precursors in the final film, this corresponds to a F127:Titania precursor ratio from 0.75 to 1.31, and this range is used as a target for initial synthesis studies. Also, the degree of orthogonal orientation is expected to increase as film thickness decreases and the critical film thickness for alignment of F127-based films will be established. A hole carrying conducting polymer (such as P3HT), when loaded into such titania film with HCP cylindrical nanopores, is hypothesized to give better photovoltaic performance as compared to a disordered or bicontinuous cubic nanopore arrangement. Confinement in cylindrical nanopores provides isolated, regioregular "wires" of conjugated polymers with tunable optoelectronic properties, which has been shown to improve hole conductivity over that in bicontinuous cubic structure. We plan to use the synthesized thin films to test this hypothesis by loading P3HT and comparing its photovoltaic performance to films with interconnected globular pores.

(153d) Modification of PCBM Crystallization Via Incorporation of C60 in Polymer/Fullerene Solar Cells

Jeff Richards, Rainie Nelson and Danilo Pozzo, Chemical Engineering, University of Washington, Seattle, WA

Polymer solar cells are a field of active research in the drive to create low-cost, flexible organic photovoltaics. Key to the development of these devices is the optimization of the active layer morphology, which has been extensively studied in a number of polymer/fullerene systems. Equally important to achieving an optimum morphology, however, is the ability to stabilize it against degradation at operating conditions that are typical of working devices. This is an issue because the continued demixing of polymer/fullerene blends leads to growth of large-scale fullerene crystallites that eventually destroy the optimum thin-film structure. This work evaluates the morphological and performance effects of the incorporation of C_{60} into blended thin-films of poly(3-hexylthiophene) and [6,6]-Phenyl C_{61} butyric acid methyl ester (PCBM). The results show that addition of C_{60} readily alters the growth-rate and morphology of PCBM crystallites under different environmental conditions. The effect of C_{60} on the growth of large PCBM crystallites is thoroughly characterized using optical microscopy, electron microscopy and UV-Visible absorption spectroscopy. Results show that C_{60} incorporation modifies fullerene aggregation and crystallization and greatly reduces the average crystal size at C_{60} loadings of ~50 wt% in the fullerene phase. Organic field-effect transistors (OFETs) and organic solar cells (OSCs) are fabricated from mixed-fullerene active layers to evaluate their performance. It is demonstrated that the use of fullerene mixtures in organic electronic applications is a viable approach to produce more stable devices and to control the growth of fullerene crystals.

(153e) Template Stripped Metal Nanostructures for Photovoltaics

Kevin M. McPeak, ETH- Zurich, Zurich, Switzerland and David J. Norris, Optical Materials Engineering Laboratory, ETH Zurich, Zurich, Switzerland

Traditional solar cell absorber materials need to be optically thick in order to completely absorb the incident radiation that is above the band gap of the absorber. This results in absorber thicknesses of a few microns for highly absorbing direct band gap materials such as CdTe to several hundred microns for weakly absorbing indirect band gap materials such as crystalline silicon. Advanced photon management via nanostructured back reflector surfaces can facilitate light trapping, anti-reflection and plasmonic effects which allow for equal or better absorption with less material. Reducing the thickness of the absorber layer is beneficial for cost reduction and cell performance since thinner absorber layers result in less bulk recombination. Reducing the thickness of the absorber layer is especially advantageous to amorphous silicon solar cells since it has been shown to mitigate the Staebler-Wronski effect.

Template stripping is a technique that relies on the weak adhesion between coinage metals (Au, Ag, Cu) and SiO_2 (e.g. native oxide covered silicon) to provide ultrasmooth metal surfaces. We report on how template stripping can be extended to fabricate flexible, wafer scale, nanostructured metal surfaces which can be incorporated as the back reflector in various solar cell devices. Furthermore, we show how template stripping is capable of producing photovoltaic back reflectors with complex, two dimensional, nanostructured gratings which result in improved device efficiency. Both theoretical and experimental results will be presented.

(153f) Nanomanufacturing of Multicomponent Plasmonic Interfaces with Broadband Solar Absorption Capability

Tao Cong, Jennifer Amey, Peter Paynter and **Radhakrishna Sureshkumar**, Department of Biomedical and Chemical Engineering, Syracuse University, Syracuse, NY

Thin-film solar cells offer much promise in affordable renewable energy harvesting: Compared to the conventional bulk solar cells which use several 100 microns thick semiconductor material, thin film technology targets the use of 1-2 micron thick layers which can be applied to hard or flexible substrates. Thin film photovoltaic cells have a potential economical advantage over expensive bulk photovoltaic cells due to the low cost, high volume scale up capabilities. However, the current conversion efficiencies of such thin films are much lower than bulk photovoltaic cells due to poor light trapping. Improving the efficiency of thin-film photovoltaics using plasmonic interfaces is an area of active research. A promising approach is the incorporation of a light-trapping layer that consists of noble metal nanoparticles onto the photovoltaic devices [1-3]. Two- or three- dimensional nanostructured plasmonic interfaces for this purpose have been fabricated by using lithography, vapor deposition and dewetting of thin metal films by ns and fs pulsed lasers [4-11]. However, economical scale up and adaptation of such processes to fabricate interfaces with multiple species/shapes/sizes in a controllable and repeatable fashion are not straightforward.

In this study, we use three methods suitable for low cost, high volume manufacturing of 2D and 3D plasmonic nanocomposites, including Layer-By-Layer (LBL) assembly on glass, spin coating nanoparticle layers on silicon and glass, and embedding nanoparticle in thin polymer films. LBL assembly provides a versatile and controllable route to immobilize functional nanoparticles onto substrates. We demonstrate that a uniform distribution of nanoparticles with controllable area coverage can be achieved using the conventional spin-coating method. Flexible thin films based on polyvinyl alcohol (PVA) with embedded multicomponent nanoparticles have also been made. The optical responses of such nanocomposites can be tuned by changing the nanoparticle type, size, shape, and composition. Applications of such plasmonic nanocomposites to solar energy harvesting, smart glasses with tunable optical properties, and antibacterial surface coating will be presented.

Acknowledgements: We acknowledge National Science Foundation grant CBET-1049454 for partial support of this research. Syracuse University has filed US and PCT patent applications based on the findings of tis work.

References:

1. H. A. Atwater and A. Polman, Nat. Mater. 9 (3), 205 (2010).

2. S. Pillai, K. R. Catchpole, T. Trupke, and M. A. Green, J. Appl. Phys. 101, 093105 (2007).

3. J. Trice, H. Garcia, R. Sureshkumar, and R. Kalyanaraman, Proc. SPIE 6648, 66480L (2007).

4. E. M. Hicks, S. Zou, G. C. Schatz, K. G. Spears, R. P. Van Duyne, L. Gunnarsson, T. Rindzevicius, B. Kasemo, and M. Käll, Nano Lett. 5 (6), 1065 (2005).

5. L. M. Campos, I. Meinel, R. G. Guino, M. Schierhorn, N. Gupta, G. D. Stucky, and C. J. Hawker, Adv. Mater. 20 (19), 3728 (2008).

6. Y. H. Lanyon, G. De Marzi, Y. E. Watson, A. J. Quinn, J. P. Gleeson, G. Redmond, and D. W. M. Arrigan, Anal. Chem. 79 (8), 3048 (2007).

7. Y. Wu, J. D. Fowlkes, P. D. Rack, J. A. Diez, and L. Kondic, Langmuir, 739 (2010).

8. L. Kondic, J. A. Diez, P. D. Rack, Y. Guan, and J. D. Fowlkes, Phys. Rev. E 79 (2), 026302 (2009).

9. C. Favazza, J. Trice, R. Kalyanaraman, and R. Sureshkumar, Appl. Phys. Lett. 91, 043105 (2007).

10. J. Trice, C. Favazza, D. Thomas, H. Garcia, R. Kalyanaraman, and R. Sureshkumar, Phys. Rev. Lett. 101 (1), 17802 (2008).

11. C. Favazza, R. Kalyanaraman, and R. Sureshkumar, Nanotechnology 17, 4229 (2006).

(153g) One-Step Metal-Assisted Chemical Etching with Silver and Copper Nanoparticles for Nanopore-Type Black Silicon Fabrication

Yen-Tien Lu, Chemical and Biomolecular Engineering, Rice University, Houston, TX and Andrew R. Barron, Chemistry, Rice University, Houston, TX

In the solar cell industry, anti-reflection (AR) coating is a popular method to suppress the reflectivity of the solar cell surface by forming destructive interference of incident light. However, AR coating only reduces the reflection for a narrow range of light wavelength and incident angle because of its functionality based on a quarter-wavelength coating. A potential replacement for the AR coating, called black silicon, has become attractive in the fields of Si solar cells and other photon devices. A graded-density surface of black silicon allows this novel Si material to possess low reflectivity and a correspondingly high absorption of visible light. Black silicon has been successfully fabricated by several different methods including reactive ion etching, femtosecond pulsed laser technique, and pulsed electrochemical etching. However, these techniques need either expensive instruments with high energy consumption or complicated fabricating processes, which make them unfavorable for industrial applications.

An economical approach for black silicon manufacturing called metal-assisted chemical etching (MacEtching) method has been developed to decrease the fabrication cost. MacEtching usually includes two steps which are metal deposition and electroless chemical etching. In the metal deposition step, novel metal nanoparticles (NPs) which can attract electrons from Si substrate are firstly deposited on a Si substrate to oxidize the underneath Si to SiO_2. Next, in the electroless chemical etching step, the as-formed SiO_2 is etched away and pits are produced under the NPs. The NPs continuously attract electrons from the Si surface to form SiO_2 and make the pits deeper and connect with each other to form black Si with Si nanowire (SiNW)-type as long as the etching is prolonged.

Based on the two-steps MacEtching technique, we have developed one-step MacEtching technique with Ag and Cu NPs to further simplify the fabrication process of black Si. In the one-step MacEtching, the metal deposition and the electroless chemical etching occur simultaneously on the Si wafer which can save time and cost of cleaning and reloading of Si wafer between the two steps. In our one-step Ag MacEtching, a mixture of $AgNO_3/H_2O_2/HF$ aqueous solution is utilized as a Si etchant under sonication. $AgNO_3$ works as an Ag NP precursor to provide Ag^+ ions for Ag NP deposition on Si wafers. H_2O_2 is a reducing agent and able to effectively reduce Ag ions to Ag NPs which oxidize Si to SiO_2. HF removes the as-formed SiO_2 underneath the Ag NPs to increase the nanopore length. We find the Ag^+ ion concentration and etching time significantly effects on the topology and reflectivity of the black silicon surface. The maximum diameter and length of obtained circular nanopore is 550 nm and 6.5 μm, respectively.

We have also studied the one-step MacEtching with Cu NPs to fabricate black silicon because the Cu precursor, $Cu(NO_3)_2$, is cheaper than Ag precursors, $AgNO_3$. Herein, H_3PO_3 is used as reducing agent instead of H
the reduction potential of H
results show that the Cu^{2+} concentrations in the etchant can not only affect the nanopore size but also the nanopore shape on the Si surface. With higher Cu^{2+} ion concentrations, the shape of the nanopores is rectangle instead of circle.

Compared to other methods, our one-step Ag and Cu-MacEtching have effectively simplified the black Si fabrication process and may cost down the cost of the facilities and energy expenditures, which is beneficial for industrial applications. On the other hand, unlike SiNW arrays, the nanopore-type black silicon possess no high-aspect-ratio needle-liked structure and are not as fragile as SiNW arrays and can better endure the stress during the solar cell assembly.

(229a) High-Performance Supercapacitor and Li Ion Battery Electrodes Using Novel Ordered Mesoporous Metal Oxide Materials

Jinwoo Lee, Department of Chemical Engineering and School of Environmental Science & Engineering, Pohang University of Science and Engineering, Pohang, South Korea

For high performance supercapacitor and Li-ion battery electrodes, various mesoporous metal oxide materials have been synthesized by employing lab-synthesized block copolymers as a structure-directing agent or ordered mesoporous silica materials as hard templates. An ordered mesoporous WO_{3-X} material with high electrical conductivity was employed for use as a supercapacitor electrode in 2.0 M H_2SO_4. The electrode made of mesoporous WO_{3-X} exhibited a high rate capability and an excellent capacitance (366 $mFcm^{-2}$, 639 Fcm^{-3}), which were attributed to the large ordered mesopores, high electrical conductivity, and high material density. Fe_3O_4 nanocrystals confined in mesocellular carbon foam (MSU-F-C) are synthesized by '*Host-Guest*' approach and tested as anode for lithium-ion batteries (LIBs). 13-27 nm magnetite Fe_3O_4 nanocrystals are formed inside ~30 nm pores of mesocellular carbon matrix. Fe_3O_4/MSU-F-C (45 wt% Fe_3O_4) exhibited 1007 mA h g^{-1} of first charge (Li de-insertion) capacity at 0.1 A g^{-1} and 111% of capacity retention at 150th cycle, and retains 37% capacity at 7 A g^{-1} (~7 C rate). For the first time, we synthesized mesostructured LTO/carbon composite (Lithium titanate/carbon) with large (> 15 nm) and uniform pores for high performance LIB anodes via a block copolymer assembly. The electrochemical characterization of this material will be presented.

(229b) 3-D Graphenic Network Composites for High-Performance Li-Ion Battery Electrodes

Xin Zhao[1], Cary M. Hayner[1], Kai Han[1], Jung Kyoo Lee[2], Mayfair C. Kung[1] and **Harold H. Kung**[1], (1)Chemical and Biological Engineering, Northwestern University, Evanston, IL, (2)Chemical Engineering, Dong-A University, Pusan, South Korea

Three-dimensional, highly conducting, networks constructed from graphene sheets have been found to be very useful scaffolds to form composite materials with high charge-storage capacities for Li ion batteries. Additional treatment of the graphene sheets to introduce in-plane defects greatly enhances the cross-plane ionic conductivity, which results in much enhanced rate capability of such materials. Such materials have exhibited power density as high as that of supercapacitors. We have constructed electrode materials suitable for use as anode or cathode for Li ion batteries. For anode materials, a three-dimensional graphenic network with in-plane defects exhibits a stable, repeatable power capability of 180 mAh/g at a rate of 2 A/g, corresponding to a recharge rate of less than 5 min. Imbedding Si nanoparticles into the network greatly improves the charge storage capacity, achieving >3100 mAh/g at low rates, and >1000 mAh/g at a high rate of 8 A/g, with a coulomb efficiency as high as 99.9%, and typically >99.6%. Cathode materials can be formed by imbedding metal fluorides or sulfides/sulfur into these scaffolds. These materials also exhibit a much higher charge storage capacity than the conventional Li metal oxides (M= Co, Ni, Mn). However, capacity fade remains an issue that needs to be overcome. The characterization results of these materials will be presented.

(229c) High Energy Density Semiconductor-Carbon Nanotube Anodes for Lithium Ion Batteries

Brian J. Landi[1], Roberta A. DiLeo[2], Melissa Thone[3], Michael W. Forney[2], Alan Raisanen[4], Matthew J. Ganter[5], Jason Staub[2] and Reginald E. Rogers Jr.[1], (1)Chemical & Biomedical Engineering, Rochester Institute of Technology, Rochester, NY, (2)NanoPower Research Laboratories, Rochester Institute of Technology, Rochester, NY, (3)Chemical & Biomedical Engineering, Rochester Institute of Technology, (4)KGCOE Semiconductor Microsystems Fabrication Laboratory, Rochester Institute of Technology, (5)Golisano Institute for Sustainability, NanoPower Research Laboratories, Rochester Institute of Technology, Rochester, NY

Alternative active materials for lithium ion batteries are under investigation to meet increasing energy storage demands. Ultra high capacity Si and Ge are promising anode materials that demonstrate capacities much higher than state-of-the-art graphite. In addition, promising developments have been made recently with electrode designs employing carbon nanotubes (CNTs) as a conductive additive or a free-standing electrode (absent of any binder or metal current collector) to serve as a physical support for silicon or germanium increasing energy and power densities. Several approaches have shown viability such as semiconductor film deposition using chemical vapor deposition (CVD) and a nanoscale hybridapproach. The performance of Si and single wall CNTs (SWCNT s) anodes, fabricated by low pressure chemical vapor deposition (LPCVD) and plasma enhanced chemical vapor deposition (PECVD), was measured to have anode extraction capacities between 1000-1500 mAh/g, even at Si weight loadings around 30%. However, dramatic differences in the rate capability and cycling for anodes from these two deposition techniques were demonstrated. Alternatively, the mixing of Ge nanoparticles (NPs) with SWCNTs to form a hybrid free-standing anode has enabled a tunable 3-dimensional electrode structure. These hybrid anodes exhibit capacities over 800 mAh/g at rates up to 1C. The balance between rate and capacity using Si and Ge has led to the development of a mixed hybrid anode that contains both materials in concert with a CNT electrode. The addition of a Si coating enhances the Ge NP anode capacity along with reducing the surface area and the first cycle loss. Full batteries incorporating hybrid anodes and conventional cathodes show potential improvements in gravimetric energy density over state-of-the-art by more than 50%.

(229d) Electrospun Nanofibers for Lithium Ion Batteries

Nathaniel S. Hansen[1,2], Daehwan Cho[2], Kyoung Woo Kim[2], Jay Hoon Park[2], Brian Williams[2] and Yong Lak Joo[2], (1)Axium Nanofibers, Ithaca, NY, (2)School of Chemical and Biomolecular Engineering, Cornell University, Ithaca, NY

Electrospinning is now a relatively common method for the production of one-dimensional nanomaterials from a scalable, cost-effective process. Recently, we have developed an approach to generate inorganic nanofibers with various nanostructures through a high-throughput, water-based solution process. We have utilized this process to generate nanofibers for lithium ion battery anode and cathode structures. It is well understood that the utilization of silicon at the anode requires nano-synthesis to overcome pulverization, and thus we have generated carbon/silicon composite anode structures that display an extremely high energy capacity (2100 mAh/g) over many cycles. Further, we display control over the carbon/silicon composite structure to minimize capacity loss for long lifetime performance. However, it is known that the generation of high energy density anode materials will not drastically improve the energy density of the battery as a whole until a high energy density cathode is synthesized. To this end, we have synthesized similar cathode materials consisting of carbon/lithium sulfide composite structures that similarly display high energy density and long lifetime. These materials, when paired together, are shown to have one of the highest theoretical energy densities thus enabling us to cost-effectively produce the highest energy density battery as a whole.

(229e) Cobalt Oxide, Multi-Wall Carbon Nanotube Polymer Network for Lithium Ion Battery Anodes

Joshua Hill[1], Yun Xu[2], Ling Fei[2], Yuling Li[1] and Hongmei Luo[3], (1)Chemical Engineering, New Mexico State University, Las Cruces, NM, (2)Chemical Engineering Department, New Mexico State University, Las cruces, NM, (3)Chemical Engineering, New Mexico State University, Las cruces, NM

Much research has been focused on the field of energy storage in lithium ion batteries. A family of cathode materials consisting of a polymer network containing cobalt oxide, and carbon nanotubes has been prepared. This combination of the polymer network with nanocrystalline cobalt oxide and multi-wall carbon nanotubes provides a novel carbon framework capable of shuttling and storage of lithium ions. Colloidal cobalt oxide was first prepared by the thermal decomposition of a metal-organic precursor. The multiwall carbon nanotubes were in used in conjunction during the synthesis of the colloidal cobalt oxide. The polymer network was prepared by mixing nanocrystalline cobalt oxide, multiwall carbon nanotube solution, with poly(ethylene)imine solution and ethylenediaminetetraacetic acid, followed by calcination at temperatures ranging from 400°C to 700°C. Characterization of the anode material was accomplished using X-ray diffraction, and scanning electron microscopy. Electrochemical properties of the materials were tested in lithium ion coin cell batteries constructed in an inert atmosphere.

(229f) Hybrid Tin Oxide Nanotube Arrays As Anodes for Lithium Ion Batteries

Umang V. Desai and Di Gao, Chemical and Petroleum Engineering, University of Pittsburgh, Pittsburgh, PA

Compared to traditional anodes based on nanostructured powdered paste, directly grown, one-dimensional ordered nanostructured array offers remarkable advantages such as fast-electron transport/collection and ion diffusion, sufficient electrochemical reaction of individual nanostructures, enhanced material-electrolyte contact area and facile accommodation of the strains caused by lithium intercalation and de-intercalation. Amongst various ordered metal oxide arrays, SnO_2 has been regarded as one of the most promising candidates for alternative anode material for Li-ion batteries due to the high theoretical capacity (781 mAh/g) compared to that of graphite (372 mAh/g). Thus, making 1D oriented nanostructure array of SnO_2 on metal substrate over a large area and further applying it as anode of battery is highly desirable for the exploration of novel electrode materials. Large-scale growth of uniform SnO_2 nanotube arrays on Ti substrate, with high orientation consistency, was achieved through a low temperature hydrothermal process. Furthermore, hybrid Sn-SnO_2 nanotubes were created by reducing the as-synthesized SnO_2 nanotubes. The performance of these arrays as anode materials in lithium-ion batteries was evaluated.

(229g) Facile Fabrication of Hollow TiO2 Nanoparticles As Safe Anode Materials for Lithium Ion Batteries

Jian Zhu, Chemical Engineering, Wayne State University, Detroit, MI, K. Y. Simon Ng, Department of Chemical Engineering and Materials Science, Wayne State University, Detroit, MI and Da Deng, Chemical Engineering and Materials Science, Wayne State University, Detroit, MI

Lithium ion batteries will play increasingly important role in expanding market of electric vehicles and green energy storage, in addition to continuously dominant role in personal electronic devices. The recall of lithium ion batteries for laptops and Chevy Volt battery fires demonstrate the critical importance of battery safety, especially for large-scale production/application in future. Safety concern requires next-generation lithium ion batteries to have carbon-alternative anode materials where its voltage plateau is higher than that of carbon to enhance safety without compromising cyclability. TiO_2 could be potentially used as such safer anode materials for lithium ion batteries, and its structural stability guarantees good cyclability in reversible lithium ion storage. In this presentation, our recent efforts on facile preparation of hollow TiO_2 nanoparticles will be reported. The hollow porous nanoparticles are formed by aggregation of smaller nanoparticles as basic building units. The hollow TiO_2 nanoparticles with high specific surface area and short diffusion path demonstrate excellent cycling performance and high current rate in reversible lithium ion storage at significantly higher voltage plateau than that of carbon.

(296a) A Nanoscale Study of the Formation of the Solid Electrolyte Interphase in Li Ion Batteries

Yanjing Chen[1,2], Mengyun Nie[3], Brett Lucht[3] and Arijit Bose[1,2], (1)Chemical Engineering, University of Rhode Island, Kingston, RI, (2)Rhode Island Consortium for Nanotechnology and Nanoscience, Kingston, RI, (3)Chemistry, University of Rhode Island, Kingston, RI

The Solid Electrolyte Interphase (SEI) is a passivating film formed on the surface of graphite anodes because of the reductive decomposition of electrolyte solvents during initial charging of a lithium ion battery (LIB). The SEI plays a crucial role on the cycling performance and storage life of LIB, protecting the electrolyte from complete decomposition but also providing additional resistance to Li ion transport. However, a comprehensive understanding of the structure and composition of the SEI is still lacking.

We have developed an in-house transmission electron microscope (TEM grid) embedded graphite/Li half cell, with a binder-free (BF) graphite anode prepared using electrophoretic deposition (EPD) of graphite nanoparticles. Using such a cell removes the 'noise' created by inactive components such as binders, and allows us to focus directly on the role of different electrolytes on the formation of the SEI and its relationship to the cell performance, and the growth of SEI at different voltages during initial charging. We were able to follow the growth of the SEI *in-situ* at the nanoscale. Coupled with X-ray photoelectron spectroscopy (XPS), Fourier transform infrared spectroscopy (FTIR) and Nuclear magnetic resonance (NMR), we have been able to identify the morphology and key chemical components of the SEI and how these vary with different electrolytes. Our results provide crucial insights into optimizing choices of electrolytes and anode materials for Li-ion batteries.

(296b) Highly Ordered Mesoporous Materials As Potential Electrodes for Li-Ion Batteries

Feng Jiao, Bryan Yonemoto and Gregory Hutchings, Chemical Engineering, University of Delaware, Newark, DE

Highly Ordered Mesoporous Materials as Potential Electrodes for Li-ion Batteries

Bryan Yonemoto, Greg Hutchings, Feng Jiao*

Center for Catalytic Science & Technology, Department of Chemical Engineering, University of Delaware, Newark, DE, 19716 USA. Fax: +1-302-831-1048; Tel: +1-302-831-3679; E-mail: jiao@udel.edu

A major goal of lithium battery research is to prepare electrodes that can sustain higher rates than present materials, for new high power applications such as HEV's. Nanomaterials support high rates, although often at the expense of volumetric energy density. Nanostructured materials can offer advantages compared with nanoparticles. For example, we have shown that mesoporous β-MnO_2 is capable to accommodate 0.92 Li^+ per Mn, while in a sharp contrast, Li^+ ion cannot intercalate into micron sized β-MnO_2. Here, we will show our recent progress in the synthesis of novel nanostructured materials as potential electrode materials for lithium-ion batteries.

There is much current interest in mesoporous metal oxides. The synthesis of such solids, especially porous materials, often requires a templating approach. Soft templates, e.g. long surfactant molecules, or hard templates, e.g. mesoporous silica, may be used. However, it is difficult to prepare metal oxides with extremely large pores (> 10 nm) by such routes. Overcoming this challenge is important if we are to prepare mesoporous transition metal oxides with enhanced electrochemical properties. We will describe the synthesis of a wide range of mesoporous transition metal oxides by using mesoporous silica as hard template. The as-prepared mesoporous compounds exhibit superior performance compared with their normal analogues.

References:

1. Ren, Y., Armstrong, A. R., Jiao, F. & Bruce, P. G. Influence of Size on the Rate of Mesoporous Electrodes for Lithium Batteries. Journal of the American Chemical Society 132, 996-1004 (2010).

2. Jiao, F., Bao, J. L., Hill, A. H. & Bruce, P. G. Synthesis of Ordered Mesoporous Li-Mn-O Spinel as a Positive Electrode for Rechargeable Lithium Batteries. Angewandte Chemie-International Edition 47, 9711-9716 (2008).

(296c) V2O5 Network Structure As Cathode for Lithium Ion Batteries

Yun Xu, Chemical Engineering Department, New Mexico State University, Las cruces, NM, Marco Dunwell, Chemical Engineering, New Mexico State University, Las Cruces, NM and **Hongmei Luo**, Chemical Engineering, New Mexico State University, Las cruces, NM

V_2O_5 has been vigorously investigated as an electrode material because of its high output voltage, and low cost, and high capacity compared to most commercial cathode materials, including $LiFePO_4$, $LiMn_2O_4$, and $LiCoO_2$. The unique crystal structure which is formed by the stacking of V_2O_5 layers makes V_2O_5 a special host for the reversible insertion and extraction of lithium ions. Many works has been devoted to improve the performance of V_2O_5, including making V_2O_5/carbon matrix composite, V_2O_5 porous structure, and nanostructures, etc.

In this work, we report a V_2O_5 network structure which is synthesized by a polymer-assisted chemical solution method. This V_2O_5 network structure delivers high and stable capacity of around 300, 200, and 120 mAh/g at current densities of 100, 400, and 800 mA/g after 50 cycles, respectively, which demonstrated an excellent stability and rate capability. The capacity and rate capability is better than most reported non-carbonaceous V_2O_5 structure. The high performance of the network structure comes from the unique two-dimensional network which bridges different length scales from the nanoscale to the micro/macro scale.

(296d) Polyaniline Nanofiber/Vanadium Pentoxide Layer-by-Layer Electrodes for Energy Storage

Lin Shao, Chemical & Environmental Engineering, Yale University, New Haven, CT, Ju-Won Jeon, Texas A&M University, College Station, TX and Jodie Lutkenhaus, Department of Chemical Engineering, Texas A&M University, College Station, TX

Li-ion batteries have garnered much interest in portable power, but in some aspects, they are limited. For example, some commercial Lithium-ion battery electrodes are subject to volumetric expansion, safety concerns, and high cost. However, nanostructured electrodes comprised of alternative materials can potentially address many of these limitations. Here, we report the layer-by-layer assembly of polyaniline (PANI) nanofibers and V_2O_5 to form hybrid electrodes for electrochemical energy storage. PANI nanofibers assist in forming a highly porous electrode (density = 0.247g/cm^3), and impart electronic conductivity. Both PANI and V_2O_5 store charge through Faradaic reactions. Together, this composite may form an electrode that surpasses either material alone. The growth of the film was investigated using profilometry and quart crystal microbalance; the thickness of these films increased linearly with respect to the number of layer pairs deposited. The non-aqueous electrochemistry of the electrode was explored using lithium as a counter and reference electrode. It was found that the electrochemical response possessed contributions from both PANI nanofibers and V_2O_5, and that the electrochemical performance of this LbL system was dependant on film thickness. A maximum capacity of 233 mAh/g at a current of 2μA/cm^2 was measured for a film of 16 layer pairs. In addition, PANI nanofibers were found to dominate the electrochromic response measured using UV-Vis spectroscopy. The composition of the film and the oxidation state of vanadium were measured and monitored using X-ray photoelectron spectroscopy (XPS). No appreciable volumetric expansion after charge/discharge cycling. Based on our current study, the formation of pernigraniline base polyaniline was attributed to the performance degradation of these films. This study indicates how LbL assembly can be used to fabricate high-performance electrodes for electrochemical energy storage.

(296e) Engineering Lixalysizo Ionic Conductive Thin Films by Atomic Layer Deposition for Lithium-Ion Battery Applications

Ya-Chuan Perng[1], **Jea Cho**[1], Daniel Membreno[2], Nick Cirigliano[2], Bruce Dunn[2] and Jane P. Chang[1], (1)Department of Chemical and Biomolecular Engineering, University of California, Los Angeles, Los Angeles, CA, (2)Department of Materials Science and Engineering, University of California, Los Angeles, Los Angeles, CA

Lithium (Li)-ion batteries have drawn much attention for their outstanding performance in portable electronics applications with the potentials to function as a power source for further minizaturized devices, including micro-systems through the utilization of 3-dimensional electrodes based on high aspect ratio pillars. To realize this potential, an ultra-thin and highly conformal electrolyte layer is needed to coat the 3D electrode array. The ionic conductor lithium aluminosilicate ($LiAlSiO_4$) synthesized by atomic layer deposition (ALD) is a promising electrolyte material for 3D battery applications not only due to its high ionic conductivity along its c-axis resulting from channels formed by the alternating tetrahedra of aluminum-oxygen (Al-O) and silicon-oxygen (Si-O), but also expected to provide similar improved cell cyclability, as reported in the preliminary studies of ultra-thin metal-oxide ALD coatings on electrodes.

The self-limiting characteristic of ALD allows for precise control of thickness and composition of complex oxides and results in a highly conformal and pinhole-free coating suitable in 3D micro-battery applications or electrolyte surface coatings. The metal precursors used in this work are tetraethyl orthosilicate (TEOS), trimethylaluminum (TMA) and lithium t-butoxide (LTB). These precursors, along with water vapor as the oxidant, were used to deposit SiO_2, Al_2O_3 and Li_2O, with the deposition rates in the range of 0.8~2Å/cycle, respectively. The deposition rate of stoichiometric $LiAlSiO_4$ was ~20Å/cycle at a temperature of 290°C. The concentration of each metal element in $Li_xAl_ySi_zO$ (LASO) thin films was found to correlate closely to ALD cycles and the associated incubation times. The crystallinity of the films after post-deposition rapid thermal annealing (RTA) was a function of cation atomic percentage. Li-ionic conductivities and the activation energy of as-deposited LASO films with respect to lithium contents as well as their relation to the film thickness were studied. The LASO ALD coating on 3D features, such as NWs and nanopaticles (NPs), were confirmed to be conformal and uniform by transmission electron microscopy (TEM) imaging. The cell performance as well as cyclability enhancement from LixAlySizO was investigated for a silicon-nanowire 3D microbattery, where SiNW was used as an anode, to explore the potentials of a solid-state SiNW battery with a solid-oxide electrolyte.

(296f) Mixed Hydrophobic Room Temperature Ionic Liquid Electrolytes for Rechargeable Lithium-Air Battery

Mahbuba Ara[1], Lixin Wang[2], Kapila Wadumesthrige[2], Steven O. Salley[1] and K. Y. Simon Ng[3], (1)Chemical Engineering and Materials Science, Wayne State University, Detroit, MI, (2)Chemical Engineering, Wayne State University, Detroit, MI, (3)Department of Chemical Engineering and Materials Science, Wayne State University, Detroit, MI

The increase in solution resistance due to the evaporation and decomposition of carbonate based electrolyte especially at high charge voltages hinders the performance (cyclability) of Lithium –Air battery. Hydrophobic room temperature ionic liquid can be used to increase the cycle numbers of lithium-air battery by eliminating the decomposition of electrolyte as well as protecting the lithium anode from hydrolysis reaction with moisture from air. Pyrrolidinium based ionic liquid N-butyl-N-methylpyrrolidinium bis (trifluoromethanesulfonyl) imide ($PYR_{14}TFSI$) shows a wide electrochemical voltage stability window. But it has high viscosity and consequently lower conductivity than standard organic electrolyte. Therefore, the mixed ionic liquid that simultaneously has the properties of high ionic conductivity and high electrochemical voltage window for battery application has been investigated. Imidazolium based ionic liquid, 1-ethyl-1-methylimidazolium bis (trifluormethylsulfonyl) imide (EMIMTFSI) has lower viscosity and higher ionic conductivity than $PYR_{14}TFSI$. This study will investigate the addition of different proportions of EMIMTFSI to $PYR_{14}TFSI$ for higher ionic conductivity as well as high electrochemical voltage stability. Optimum mixture of the ionic liquids will be studied for the long term cyclic performances of lithium-air battery. Different concentrations of lithium salt (Lithium bis (trifluoromethanesulfonyl) imide (LiTFSI)) to the optimum mixture will be examined for the best performance of lithium-air battery. The optimum ionic liquid electrolyte will be used in conjunction with perovskite type bifunctional catalyst for enhanced capacity and rechargeability of lithium-air battery.

(296g) Novel Graphene Supported Bifunctional Catalysts Combined with Ionic Liquid Electrolyte in Rechargeable Lithium-Air Batteries for Prolong Cyclic Stability

Lixin Wang[1], Mahbuba Ara[2], Steven O. Salley[2] and K. Y. Simon Ng[3], (1)Chemical Engineering, Wayne State University, Detroit, MI, (2)Chemical Engineering and Materials Science, Wayne State University, Detroit, MI, (3)Department of Chemical Engineering and Material Science, Wayne State University, Detroit, MI

Li-air battery is widely considered to be the next generation battery chemistry with extremely high energy density comparable to gasoline. They are very promising technology to prolong the driving ranges of electric vehicles. Compare to their high theoretical value, the practical energy density of Li-air battery is much lower due to the slow kinetics of both discharge and charge reactions and hence, high overpotential at air cathode. To address this technical barrier, Pt catalyst has been used to overcome the activation energy and consequently reduce the overpotential. However, the high cost of Pt still restricts the wide application of Pt in air-cathode. In this presentation, low cost, bi-functional catalyst has been developed to replace Pt catalyst. A Li-Air system capable of 100 cycles with an average capacity of 2000 mAh/g and total energy efficiency of 75% has been achieved compared to 30 cycles with 1800 mAh/g average energy density and 85% energy efficiency as a control.

Another important factor that hinders the cycle life of Li-air battery is the increasing solution resistance due to the consumption of carbonate based electrolyte, especially during the high charge voltages. Electrolyte oxidation, evaporation and side reactions with active electrode materials are believed to be the main reason of the electrolyte lost. In this presentation, the conductivity, voltage window, viscosity and other electrochemical related properties of several ionic liquids as well as their mixture at different mass ration have been also investigated. by employing the optimized ionic liquid electrolyte, the high resistance after cycling is expected to be addressed and the cycle life of Li-air batteries are expected to be prolonged to 300 cycles.

(363a) Rapid Fabrication of Binder-Free Multilayered Electrodes of Polyaniline Nanofibers and Multi-Walled Carbon Nanotubes

MD Nasim Hyder[1], Yang Shao-Horn[2] and Paula T. Hammond[1], (1)Chemical Engineering, Massachusetts Institute of Technology, Cambridge, MA, (2)Mechanical Engineering, Massachusetts Institute of Technology, Cambridge, MA

Next generations electrochemical energy storage devices require high-power, high-energy density, simultaneously. To achieve this goal, it is essential to identify novel electrode materials based on nanoparticles, nanotubes or nanofibers with decreased transport length of ions/electrons and high surface area. Nanomaterials such as carbon nanotubes and conducting polymeric nanofibers /nanoparticles with their exotic properties can significantly increase power (W/kg) and energy density (Wh/kg) in addition, to their longer life cycles and charging-discharging efficiency. We assembled free-standing multilayered thick (10 to 60 microns) electrodes of conducting polyaniline (PANi) and multiwalled carbon nanotubes (MWNTs) that are robust, binder-free using rapid spray and vacuum-assisted layer by layer techniques. The electrode microstructure reveals porous nanostructure with high surface area; Preliminary results for these highly conductive electrodes (4.4 S/cm) show high pseudocapacitance of 210 F/cm^3 and are stable over 1000 cycles. We will present our findings on the optimization of the free standing asymmetric electrode morphology and electrochemical performance in aqueous and organic electrolytes.

(363b) Tunable 3-D MnO2/Graphene Composites for Next-Generation High Energy Density Supercapacitors

Lixin Wang, Chemical Engineering, Wayne State University, Detroit, MI, Da Deng, Chemical Engineering and Materials Science, Wayne State University, Detroit, MI and K. Y. Simon Ng, Department of Chemical Engineering and Material Science, Wayne State University, Detroit, MI

Supercapacitors are promising electrochemical devices for energy storage. Supercapacitors, with advantages of high power density, excellent cyclability, high efficiency and low fabrication cost, can find important applications in different areas, such as electric vehicles, industrial equipment, intermittent energy storage and electronic devices. However, current electric double-layer capacitors have low energy density, which restrict the even broader applications of supercapacitors. In the last decade, significant amount of effort has been invested on increasing the energy density of supercapacitors, specifically, the introduction of nanomaterials as electrode materials with large surface area and enhanced conductivity has achieved improvement.

In this presentation, we will present our recent effort on the design and fabrication of a high-energy 3-D composite of 2-D graphene nano-sheets and 1-D MnO2 nano-wires where graphene nano-sheets provide high surface area, high conductivity as well as excellent mechanical support for high-capacitance MnO2. MnO2 serves as spacer to prevent graphene nano sheet from restacking and also provides pseudo capacitance to extent energy performance.

The 3-D composites were fabricated by facile hydrothermal method followed by modified Hummus method. The structure and composition of the 3-D composites can be easily tuned by adjusting the reaction time, temperature, concentration of precursors. The size, morphology and crystallization of MnO2 nanowires and the mass ratio between MnO2 and graphene were optimized to achieve highly improved electrochemical performance, which demonstrated the potential applications of the 3-D composite for next-generation high-energy density supercapacitors. Three electrodes and two electrodes set up were employed to test electrochemical performance of the as prepared 3-D MnO2 and graphene composites and consistent results were obtained.

(363c) The Electrocapacitive Properties of Organometallic Polymer/Multiwalled Carbon Nanotube Hybrids

Xianwen Mao, Demetra Achilleos, Fritz Simeon, Gregory C. Rutledge and T. Alan Hatton, Chemical Engineering, Massachusetts Institute of Technology, Cambridge, MA

The Electrocapacitive Properties of Organometallic Polymer/Multiwalled Carbon Nanotube Hybrids

*Xianwen Mao, Demetra S. Achilleos, Fritz Simeon, Gregory C. Rutledge and T. Alan Hatton**

There is growing interest in adapting redox-active materials for supercapacitor applications, owing to expected synergistic effects when the capacitive charging of the electrochemical double layer can be combined with a redox (pseudocapacitive) reaction. When both charge storage mechanisms work in parallel, the overall capacitance of the composite electrode becomes the sum of the electrochemical double layer capacitance and the redox capacity. Various redox polymer electrodes have been developed for this application, but they generally suffer from limited chemical reversibility and lack long-term cycling stability in aqueous electrolyte solutions.

The redox responsiveness of hybrids comprised of organometallic polymers and multiwalled carbon nanotubes (MWCNT) makes it possible to develop supercapacitors with the desired redox capacity. We have optimized the composition of the polymer/MWCNT hybrids to achieve enhanced performance and stability as electrochemical capacitors. In this paper, we report on the morphology and structure of the hybrids as investigated by transmission electron microscopy, scanning electron microscopy and nitrogen adsorption/desorption isotherms, and study the capacitive performance, rate capability and long-term cycling stability of the hybrids using cyclic voltammetry and galvanostatic charge/discharge techniques. The charge-storage mechanism is examined based on the results of electrochemical impedance spectroscopy.

(363d) Activated Carbons Derived From Orange Peel Wastes As Electrode Materials In Electrochemical Double Layer Capacitor

Arenst Andreas Arie, Chemical Engineering, Parahyangan University, Bandung, Indonesia and Joong Kee Lee, Clean Energy and Chemical Engineering, University of Science and Technology, Daejeon, South Korea

As an electrical energy storage device, electrochemical double layer capacitor (EDLC) finds various applications in electronic products and alternative power source due to its higher energy density, fast discharge/charge time, safety and long-term operation stability. In this work, orange peel was utilized as the precursor for activated electrodes in EDLC. The surface of the carbon material was treated with the oxidative chemical agents, $ZnCl_2$. The surface morphology, surface composition, and surface area of the as-prepared carbon materials were observed by scanning electron microscope, X-ray photoelectron spectroscopy and Brunauer–Emmett–Teller Analysis, respectively. Electrochemical characteristics of the carbon sample were studied by cyclic voltammetry, galvanostatic charge/discharge, and electrochemical impedance spectroscopy measurements in $LiPF_6$ aqueous solution. Compared with the non-activated carbon, activated carbon showed superior capacitive performance. The activation carbon presented a high specific gravimetric capacitance of 180 F g^{-1}. The good electrochemical performance of the activated carbon was attributed to the high surface area and larger pore volume.

(363e) Lignin Utilization in Production of Electrospun Carbon Nanofibers with Potential for Energy Storage Applications

Chuilin Lai[1], Lifeng Zhang[1], Hao Fong[1] and Lew P. Christopher[2], (1)Chemistry, South Dakota School of Mines & Technology, Rapid City, SD, (2)Center for Bioprocessing Research & Development, South Dakota School of Mines & Technology, Rapid City, SD

Lignin, a naturally existing biopolymer contained in lignocellulosic biomass at 20-30% on average, is a high-volume byproduct of wood pulping which in the U.S. alone exceeds 15 million tons per year. Nonetheless, lignin has been utilized mainly as a low-grade fuel for production of heat and power to date. There is however a growing demand to identify applications with higher economic value for lignin. Although production of carbon fibers from lignin has been reported in literature, the low mechanical properties obtained so far have prevented their commercial use in reinforcement and/or composites applications. Nano-fibrous felts consisting of porous carbon nanofibers derived from low-cost lignin, however, have potential applications that rely on large specific surface area such as catalyst support, supercapacitor electrode, and hydrogen storage materials. In this research, innovative carbonaceous nano-fibrous felts with high specific surface area were developed by electrospinning of aqueous mixtures of alkali lignin and poly(vinyl alcohol) with varied mass ratios followed by heat treatments for stabilization and carbonization. The developed carbonaceous nano-fibrous felts had up to 19.4-fold larger pore volume and 7-fold greater specific surface area than the PVA-based control prepared without use of lignin, and are therefore expected to find important commercial applications.

(363f) In-Situ Characterization of Transition Metal Nitrides for Supercapacitor Electrodes

Priyanka Pande[1], Alice E. S. Sleightholme[1], Paul G Rasmussen[1], Aniruddha Deb[2], James Penner-Hahn[2] and Levi T. Thompson[1], (1)Department of Chemical Engineering, University of Michigan, Ann Arbor, MI, (2)Department of Chemistry, University of Michigan, Ann Arbor, MI

In-situ Characterization of Transition Metal Nitrides for Supercapacitor Electrodes

Priyanka Pande[a], Alice E. S. Sleightholme[a], Paul Rasmussen[ac], Aniruddha Deb[b], James Penner-Hahn[b],

Levi T Thompson*[ac]

[a] Department of Chemical Engineering

[b] Department of Chemistry

[c] Hydrogen Energy Technology Laboratory

University of Michigan, Ann Arbor, MI- 48109-2100.

E-mail: ltt@umich.edu

Early transition metal nitrides are promising candidates for use in supercapacitors due to their high electronic conductivities, surface areas (up to 200 m^2/g) and electrochemical stabilities [1,2]. Of these, V and Mo nitrides have been demonstrated to possess the highest capacitances. For example, VN with 1340 Fg^{-1} in aqueous KOH [3] and γ-Mo_2N with 380 Fg^{-1} in aqueous H_2SO_4 [4] have been reported. Further development of these materials would benefit from a better understanding of their charge storage mechanisms. In this paper we identify active species on the electrode-surface during charge-storage, present results from *in-situ* x-ray absorption spectroscopy (XAS), and suggest mechanisms that reconcile these results. Additionally we will present results from impedance spectroscopy and charge-discharge, indicating the device level performance for these materials.

The nanostructured V and Mo nitrides were synthesized via temperature-programmed-reaction of their oxide precursors with anhydrous NH_3 followed by passivation in 1% O_2/He at room temperature to form a oxygen-rich passivation layer preventing bulk oxidation on exposure to air [1]. Physical characterization was performed using BET surface area analysis and X-ray diffraction. Our previous results using electrolyte ion-isolation techniques suggested that OH^- and H^+ were the contributing species in charge storage for VN and γ-Mo_2N [4]. Chronopotentiometry at varying concentrations of the electrolyte was used to measure the open circuit potential (OCP). The OCP values were then used to establish the relationship between the charge-transferred and the active species. *In-situ* XAS experiments were carried out in the cell shown in Figure 1. The stability range was determined using cyclic voltammetry for VN in 0.1M KOH and γ-Mo_2N in 0.1M H_2SO_4 inside the XAS cell (*vs* a Pt wire reference electrode). The stable potential range was divided into potential steps and chronoamperometry was carried out while x-ray absorption spectra were collected at these voltages from highest to lowest potential and then in reverse.

The chronopotentiometry results for VN and γ-Mo_2N suggested that 0.5 and 1.7 electrons are transferred per OH^- and H^+ reacted, respectively. The *in-situ* x-ray absorption near-edge spectra (XANES) for γ-Mo_2N are shown in Figure 2. Based on the Mo-edge shift collected at the various potentials, we estimated the change in oxidation state for Mo at these voltages (Figure 3). These results indicated the following changes in the Mo oxidation state during charge-storage: $Mo^{3.6+}$ $Mo^{3.2+}$ $Mo^{2.8+}$ $Mo^{2.4+}$. The results for forward and reverse scans revealed minimal change in Mo oxidation state at a given potential, indicating the reversibility and cyclability of the material. *In-situ* X-ray absorption fine structure (XAFS) results will also be presented to show the changes in the coordination sphere. Parallel experiments with VN required recording spectra in fluorescence mode instead of the transmission mode as was used for γ-Mo_2N. The fluorescence mode is more difficult to analyze, but our preliminary interpretations indicate reversible redox reactions for VN.

Reference:

[1] Cladridge J B, York A P E, Brungs A J, Green Malcolm L H, Chem. Mater.12 (2000) 132.

[2] Wixom M R, Tarnowski D J, Parker J M, Lee J Q, Chen P -L, Song I, Thompson L T, Mat. Res. Soc. Symp. Proc. 496 (1998) 643.

[3] Choi D, Kumta P N, Electrochem. Solid-State Lett. 8 8 (2005) A418.

[4] Pande P, Rasmussen P G, Thompson L T, J. Power Sources. 207 (2012) 212.

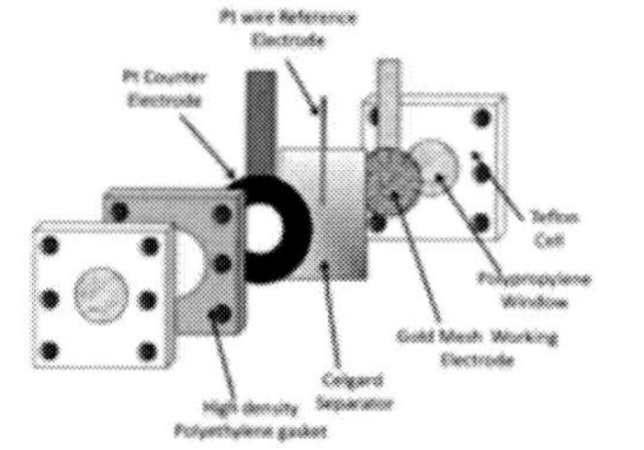

Figure 1: Schematic of *in-situ* x-ray absorption spectroscopy cell.

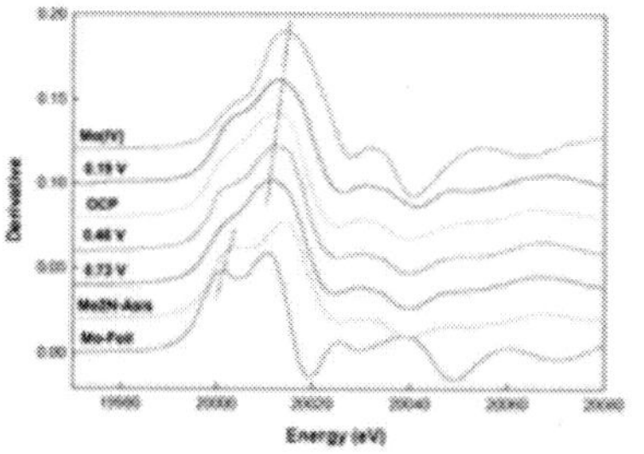

Figure 2: *In-situ* XANES spectra of γ-Mo_2N in 0.1M H_2SO_4 at various potential steps in transmission mode. Mo_2N As-is is the dry electrode. Mo-foil and Mo (IV) were used as model compounds.

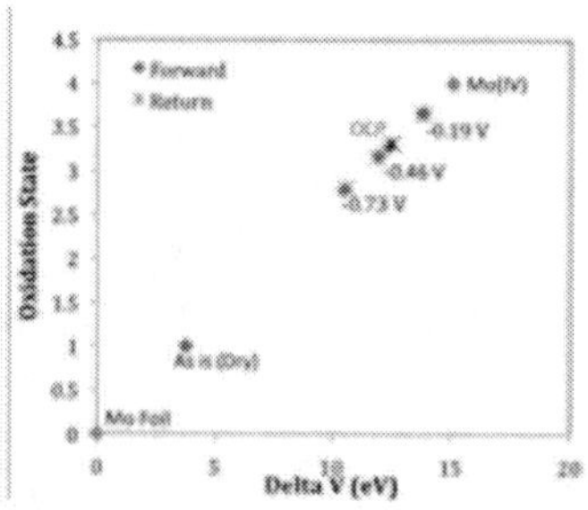

Figure 3: Changes in oxidation state of γ-Mo_2N in 0.1 M H_2SO_4 at various potentials during forward and reverse scans.

(363g) Nanostructured Vanadium Nitride for Supercapacitor Applications

Prashanth Jampani Hanumantha[1], Moni Kanchan Datta[2], Karan Kadakia[1], Dae Ho Hong[3], Michael C. Tam[1], Ayyakkannu Manivannan[4] and Prashant Kumta[1,2,5], (1)Chemical and Petroleum Engineering, University of Pittsburgh, Pittsburgh, PA, (2)Bioengineering, University of Pittsburgh, Pittsburgh, PA, (3)BioEngineering, University of Pittsburgh, Pittsburgh, PA, (4)National Energy Technology Laboratory, (5)Mechanical Engineering and Materials Science, University of Pittsburgh, Pittsburgh, PA

There is a dire need to provide a sustainable solution to the growing global demand for energy. It is this incessant demand that is the primary driver fueling various new current and emerging technologies including lithium ion batteries, supercapacitors and fuel cells. Supercapacitors hold a unique place among these energy storage devices on account of their high power density and long cyclability. Supercapacitors are electrochemical charge storage devices capable of delivering large bursts of charge over a very short period of time. It is due to this unique capability that supercapacitors occupy a very important role in the growing alternative energy field. High surface area carbons, hydrated ruthenium oxide and $MnO_2$are among the primary materials being considered for supercapacitor applications. Our group has previously identified vanadium nitride (VN) as a promising electrode material for supercapacitors on account of its high energy density with capacitances as high as 1300 F/g.

In this work, we present a facile one-step direct mechanical milling approach for the synthesis of nanostructured vanadium nitride (VN) particles. The electrochemical mechanism and material properties influencing the supercapacitor properties of VN have been explored. The surface properties of VN have also been studied using X-ray photo-electron spectroscopy (XPS) and correlated to the electrochemical characteristics. Stable capacitances of upto 100F/g are reported here for VN prepared using High Energy Mechanical Milling (HEMM). Degradation of the capacitance upon cycling is studied and possible mechanisms for the same have been proposed. Results of these studies will be presented and discussed.

(438a) Design Principles of Telluride-Based Nanowire Heterostructures for Thermoelectric Applications

Yue Wu, School of Chemical Engineering, Purdue University, West Lafayette, IN

Thermoelectric (TE) devices, which can perform a direct conversion between thermal and electrical energy, have attracted great attention due to their promising potential in improving the energy efficiency and in solid-state cooling. However, the low efficiency of the TE materials prohibits their wide applications. Certain TE materials, such as Bi_2Te_3/Sb_2Te_3 superlattice film (*ZT*~2.4) and $AgPb_mSbTe_{2+m}$ bulk crystals (*ZT*~2.2), although possessing high performance due to the improved phonon scattering at nanoscale interfaces and grain boundaries, require very complicated material composition or extremely expensive/time-consuming manufacture process such as molecular beam epitaxy. Theoretical predictions and initial experimental results have suggested that one-dimensional (1D) nanostructures, especially the nanowire heterostructures, which take the advantages of both quantum confinement to enhance the power factor and phonon scattering at nanowire surface and compositional interfaces to lower thermal conductivity, could offer a much higher *ZT* value. Meanwhile, the syntheses of various 1D nanowire heterostructures have been demonstrated through the chemical vapor deposition process based on vapor-liquid-solid (VLS) growth mechanism as well as the pulsed electrodeposition, but it is still a great challenge to obtain high-quality thermoelectric nanowire heterostuctures at a simple yet scalable way. Here we present a design principle to develop new categories of telluride-based thermoelectric nanowire heterostructures through rational solution-phase reactions. The synthesis yields "barbell" and "dumbbell" compositional-modulated nanowire heterostructures with narrow diameter and length distribution. Initial characterizations of the hot-pressed nanostructured bulk pellets of these heterostructure show a largely enhanced Seebeck coefficient and greatly reduced thermal conductivity, which lead to an improved thermoelectric figure of merit. This approach opens up new platforms to investigate the phonon scattering and energy filtering.

(438b) Excess Thermopower in Carbon Nanotubes Using Thermopower Waves

Joel T. Abrahamson, Bernat Sempere, Michael P. Walsh, Jared M. Forman, Sayalee G. Mahajan and Michael S. Strano, Department of Chemical Engineering, Massachusetts Institute of Technology, Cambridge, MA

The nonlinear coupling between an exothermic chemical reaction and a nanowire or nanotube with large axial heat conduction results in a self-propagating thermal wave guided along the nano-conduit. The reaction wave induces a concomitant thermopower wave of high power density, resulting in an electrical current in the same direction. Thermopower waves also tend to produce monopolar voltage pulses, although conventional thermoelectric theory clearly predicts bipolar voltage, and they propagate in materials with both high electrical and thermal conductivity. Temperature and voltage measurements of thermopower waves on single-walled carbon nanotubes (SWNTs) show that they can generate voltage and power in excess of the predictions of the Seebeck effect for the temperature gradient, as much as four times greater power. We hypothesize that the excess thermopower stems from a chemical potential gradient across the SWNTs with a new model that describes the electrical properties of the waves. The fuel (prototypically picramide) and ambient oxygen adsorb and dope the SWNTs ahead of the wave, and desorb and react in the hot zone behind the wave front. Raman spectroscopy confirms that picramide dopes SWNTs. Furthermore, the excess thermopower depends on the mass of fuel added (relative to SWNT mass), and the chemical potential difference matches the magnitude of the excess thermopower. This model relies on SWNT property data measured in the literature rather than fit parameters. Thermopower waves thus provide an exciting new class of highly scalable, stable energy storage devices, and offer the possibility of greatly exceeding the power limitations of conventional thermoelectric devices.

(438c) First Principles Model for Predicting the Thermoelectric Properties of Complex Antimonides and Oxides

Maria Stoica, Alireza Faghaninia, Zhou Yu and Cynthia S. Lo, Energy, Environmental and Chemical Engineering, Washington University in St. Louis, St. Louis, MO

Nanostructured inorganic materials have in the past ten years seen a rise in interest for thermoelectric applications, due to the enhanced performance exhibited at low dimensions. In our work, we study the effect of dimensionality on the thermoelectric properties of Co- and Mo- based antimonides and doped oxides. Specifically, we calculate the Seebeck coefficient, "dark" electrical conductivity, and thermal conductivity of these materials. We first perform Molecular Dynamics (MD) simulations and Density Functional Theory (DFT) calculations, where we perform Γ-point calculations as well as integrating over the first Brillouin zone. We then perform post-processing of the momentum/optical matrix elements to calculate the desired properties and predict the behavior of both known (e.g., unfilled bulk) and designed (e.g., filled bulk, nanostructured) materials. For the CoX_3 compounds (X=P, As, Sb), we look at Ag- and Au-filled skutterudites; we also compare their performance to Mo-based antimonides. For the oxides, we have performed transition metal doping of ZnO and $CuAlO_2$ compounds with oxygen vacancies, as well as direct and charge-compensated replacement of dopant atoms. We also perform computations for both bulk and 2-D nanowires for each group of materials, and demonstrate the enhancement of thermoelectric properties as a result of nanostructuring.

(438d) Characterizing Thermal Transport in Self Assembled Monolayer Junctions

Shubhaditya Majumdar[1], Scott N. Schiffres[1], Alan J. H. McGaughey[1] and Jonathan A. Malen[2], (1)Mechanical Engineering, Carnegie Mellon University, Pittsburgh, PA, (2)Department of Mechanical Engineering, Carnegie Mellon University, Pittsburgh, PA

Our investigation involves studies of the thermal transport properties of a single organic-inorganic interface – a self-assembled monolayer (SAM) junction. Molecular junctions have generated excitement as potential forerunners in organic electronics and are fundamental to the development of new hybrid materials. Characterization of their thermal properties is important in understanding how they will dissipate heat, a parasitic byproduct in their application in electronics and energy conversion. Accurate quantification of their thermal conductance, which describes this behavior, is thus essential.

Complementary experimental and computational approaches will address a prototypical Au-SAM-Au interface, where an alkane based SAM is initially considered. Molecular dynamics simulations are being implemented to model and study the effect of molecule length, choice of leads and other factors affecting bonding between the organic and inorganic components. Independent experimental measurements on transfer printed leads on SAMs are being performed using frequency domain thermoreflectance – a laser-based non-contact measurement technique. Initial measurements indicate that the Au-SAM-Au interface has a higher conductance than previously studied SAM interfaces, perhaps because prior measurements considered SAMs sandwiched between dissimilar contacts. Hence, this work is novel because the contacts are identical, and the interface conductance can be directly related to the SAM properties. This study is thus expected to provide a detailed picture on the thermal transport phenomenon across such interfaces.

(438e) Studies of Thermal Conductivity in Hybrid Organic-Inorganic Nanocrystal Superlattices

Wee-Liat Ong[1], Sara Rupich[2], Dmitri Talapin[2], Alan J. H. McGaughey[1] and Jonathan A. Malen[1], (1)Mechanical Engineering, Carnegie Mellon University, Pittsburgh, PA, (2)Chemistry, University of Chicago

Nanocrystal superlattices (NCSLs) are 3-D arrays of inorganic spheres bridged by self-assembled molecular chains. NCSL thin films hold promise for thermal management and energy conversion applications as studies have shown that their electronic and thermal transport properties can be decoupled. This unique feature allows them to be tailored according to the specific needs of the applications (e.g. high figures of merit materials for thermoelectric, high-efficiency photovoltaic materials). An additional advantage of this material comes from their solution-based fabrication technique which is readily scaled up for large volume and cheaper than conventional nanostructuring methods.

Electronic transport in NCSLs has been studied extensively. Little is known, however, about the nature of thermal transport in these complicated materials. Both experimental measurements and preliminary modelling work will be presented to elucidate the thermal transport mechanisms in NCSLs. Several factors affecting the NCSL thermal conductivity including the geometry and chemical compositions of the NCSL, will be shown.

(511a) Electrochemical Reduced Graphene Oxide-Nickel Nanocomposites for Hydrogen Production From Urea Electrolysis

Dan Wang, Wei Yan, Santosh H. Vijapur and Gerardine G. Botte, Chemical and Biomolecular Engineering Department, Center for Electrochemical Engineering Research, Ohio University, Athens, OH

Urea ($(NH_2)_2CO$) has attracted attention as a hydrogen carrier for long-term sustainable energy supply. In comparison with other gas/liquid hydrogen carriers, such as methanol, ethanol, ammonia, and so on, urea is a stable and non-flammable solid, which make it easy for hydrogen storage and transportation. Effective catalysts and methods for urea decomposition are therefore necessary for hydrogen production from urea. Recently, electrocatalytic oxidation with inexpensive bulk nickel catalyst was introduced as an approach for urea decomposition and hydrogen production. On the other hand, graphene, a two-dimensional (2D) monolayer of sp^2-hybridized carbon atoms arranged in a honeycomb network, exhibits excellent mechanical, thermal, electrical and optical properties, large specific surface area and chemical stability. These superb properties offer graphene many potential applications in nanoelectronics, batteries, fuel cells, electrocatalysis, and so on.

Within this context, this paper focuses on the synthesis of graphene-nickel nanocomposites through a one-step electrochemical reduction process and the development of nanocomposites as electrocatalysts for hydrogen production via urea electrolysis. Atomic force microscopy (AFM), Scanning electron microscopy (SEM), UV-Vis absorption, and Raman spectroscopy are used to characterize the synthesized electrochemical reduced graphene oxide-nickel nanocomposites. Cyclic voltammetry (CV) and polarization techniques are used to investigate and evaluate the performance of the synthesized nanocomposites for urea electrolysis.

(511b) Electrospun Ceria-Based Water Gas Shift Catalyst Mats: Preparation, Characterization, and Integration Into a Membrane Reactor

William T. Gibbons, Chemical and Biomolecular Engineering, University of Maryland, College Park, MD and Gregory S. Jackson, Mechanical Engineering, University of Maryland, College Park, MD

Catalytically active non-woven ceramic fiber mats (with fibers ~200 nm dia) containing palladium and copper were prepared via electrospinning. The catalysts were screened for activity and stability under simple WGS conditions (5% CO, 10% H_2O, balance Ar). Varied preparation of the spinning solution was shown to affect both the formation of the final fiber mat as well as catalyst activity and stability. Additionally, mat formation and structure depends strongly on the polymer removal step during synthesis, and careful polymer removal can produce a strong and stable mat of material. Such material is easily integrated into a planar membrane reactor, which uses a 60/40 wt% Pd/Cu foil as the H_2 selective layer. Testing of the membrane reactor indicates a substantial CO conversion increase in the presence of the membrane, and thus shows the importance of process intensification. Long-term testing of the catalysts show gradual activity loss thought to be due in part to reduction of the ceria. This work shows the ease with which fiber-based catalyst materials can be integrated into systems.

Studies on Nanostructured Bilayered Semiconductor System in PEC Water Splitting for Hydrogen Production

Surbhi Choudhary[1], *Sumant Upadhyay*[1], *Pushpendra Kumar*[1], *Nirupama Singh*[1], *Vibha R. Satsangi*[2], *Rohit Shrivastav*[1] *and Sahab Dass*[1]

[1]*Department of Chemistry, Dayalbagh Educational Institute, Agra (India)*
[2]*Department of Physics & Computer Sciences, Dayalbagh Educational Institute, Agra (India)*

Abstract: Performance of photo-electrodes in photoelectrochemical (PEC) water splitting for hydrogen production is expected to improve by fabricating bilayered electrodes consisting of two semiconductors possessing different energy levels for their corresponding conduction and valence band. The small/mid band gap semiconductor is primarily responsible for visible light absorption and sensitizing the large band gap semiconductor through electron and/or hole injection. The energy layers in the bilayered semiconductors can cover visible spectrum thereby offering synergistic effect and better output is achieved. In the present study, an attempt has been made to synthesize bilayered $SrTiO_3$ (~3.4 eV) / WO_3 (~2.8eV) thin films by sol-gel spin-coating process on conducting glass substrate to explore its application in PEC cell. Maximum photocurrent density 0.091 mA/cm^2 at 0 V/SCE was observed for bilayered $SrTiO_3/WO_3$ thin film photoelectrode which was two folds higher than offered by $SrTiO_3$ (0.046 mA/cm^2) and much higher than that for WO_3 thin films. The enhanced photoelectrochemical behaviour of bilayered photoelectrodes can be attributed to: (a) the extended absorption towards visible region, (b) formation of mixed oxide at the interface and (c) enhanced charge separation of both the materials resulting in decrease of recombination of charge carriers.

Introduction

Hydrogen is considered as non-polluting, clean and inexhaustible energy carrier for the future due to its unique physical and chemical properties [1]. Photoelectrochemical generation of hydrogen has been the subject of intensive research as it offers direct conversion of solar energy into chemical energy. In PEC water splitting, photon absorption leads to electron-hole pair generation, separation and transfer at the relative electrode to split water into hydrogen and oxygen [2,3]. Single semiconductor material cannot satisfy the basic prerequisties of efficient PEC water splitting [4]. Therefore a new approach has been introduced in which inexpensive and stable oxide photoanodes following combination of two different bandgap in a single photoelectrode can be used to afford reasonable solar water splitting efficiency. The advantage of bilayered structures is that both semiconductor simultaneously fulfill the energetic requirement and can absorb wide region of the solar spectrum. It has been reported that n/n semiconductor heterojunction thin films have different band bending properties near the junction and have shown more promising performance in PEC cell [4].

In the present work, nanostructured $SrTiO_3/WO_3$ was fabricated by sol-gel method and its PEC performance was studied with respect to PEC performance offered by $SrTiO_3$ and WO_3 films individually.

Experimental Details

Tungsten oxide precursor was prepared by dissolving ammonium tungstate in glacial acetic acid aqueous solution in 3 : 2 volume ratio. After ultrasonication for 15 min, 5 ml H_2O_2 was added slowly into the as-prepared solution, stirred continuously for 8 h at room temperature leading to the formation of yellow coloured solution of WO_3. Similarly precursor solution of $SrTiO_3$ thin film was prepared by mixing two solutions: i) strontium acetate in glacial acetic acid and ii) titanium tetra isopropoxide in 2- methoxyethanol. The resultant solution was then stirred for 45 min to prepare the stoichiometric, transparent, and stable strontium titanate precursor. The precursor solution of WO_3 was spin coated on ITO (SnO_2:In) substrate at 2500 rpm for 20 seconds followed by drying at 80°C for 10 min. The thickness of film was maintained by successive deposition of one layer over the previously deposited layer. Finally the films were sintered at 450°C for 2 h in order to obtain homogenous and pale yellow coloured WO_3 thin films. In order to prepare $SrTiO_3/WO_3$ thin films, $SrTiO_3$ precursor solution was spin coated at 2000 rpm for 20 seconds over as-prepared WO_3 thin films. These films were then dried at 80°C for 10 min and after attaining the desired thickness by controlling the number of layers, sintered at 600°C for an hour. $SrTiO_3$, WO_3 and all the $SrTiO_3/WO_3$ bilayered films of varying thicknessess [Table 1] were subjected to various characterization viz. XRD analysis, spectrometric measurement and SEM analysis for the determination of crystalline phase, absorption band energy and surface morphology respectively. These thin films are converted into photoelectrode with the assistance of copper wire and silver paste (Hysol) and employed as working electrode alongwith counter (platinum) electrode and reference (standard calomel) electrode in photoelectrochemical cell.

Table 1: Sample description and Photoelectrochemical details of various samples.

Sample ID	Sample details	Film Thickness(nm)			Open Circuit Voltage V/SCE	Photocurrent Density at zero bias ($mAcm^{-2}$)
		$SrTiO_3$	WO_3	Overall Thickness		
ST	$SrTiO_3$	130	-	130	-0.858	0.046
W	WO_3	-	242	242	-0.711	0.0002
L_1	$SrTiO_3(1)/WO_3(3)$	65	242	307	-0.910	0.067
L_2	$SrTiO_3(2)/WO_3(3)$	130	242	372	-0.994	0.091
L_3	$SrTiO_3(3)/WO_3(3)$	195	242	437	-0.940	0.076

Results and Discussion

XRD pattern of $SrTiO_3$, WO_3 and all the $SrTiO_3/WO_3$ bilayered thin films has been shown in Fig.1. The observed diffraction peaks in the recorded XRD pattern matches well with the values available in the JCPDS card no. 01-075-2187 and 00-35-0734 for WO_3 and $SrTiO_3$ respectively. WO_3 thin film show diffraction peaks at the 2θ angle 23.10, 24.12, 28.80, 33.67, 34.05 and 47.27° correspond to (001), (200), (111), (201), (220) and (002) planes of orthorhombic WO_3 phase, respectively, while $SrTiO_3$ thin film exhibited peaks at 2θ = 32.42, 39.98, 46.48 and 57.79 corresponding to the (110), (111), (200) and (211) reflections confirming the formation of cubic strontium titanate. All bilayered $SrTiO_3/WO_3$ thin film samples persists less intensity peaks of $SrTiO_3$ and WO_3, alongwith dominant 2θ peaks at 27.65, 33.10, 45.14, 47.43 and 57.09 corresponding to the (112), (200), (204), (220) and (224) planes of strontium tungsten oxide ($SrWO_4$). The similar ionic radii of W^{+6} (0.62 Å) [5] and Ti^{+4} ions (0.605 Å) [6] facilitates easy substitution of ions at the interface resulting into formation of mixed oxide viz. $SrWO_4$, in all bilayered $SrTiO_3/WO_3$ thin films. The average crystallite size of WO_3 and $SrTiO_3$ are 39 nm and 28 nm respectively as calculated by Debye-Scherrer's equation [7].

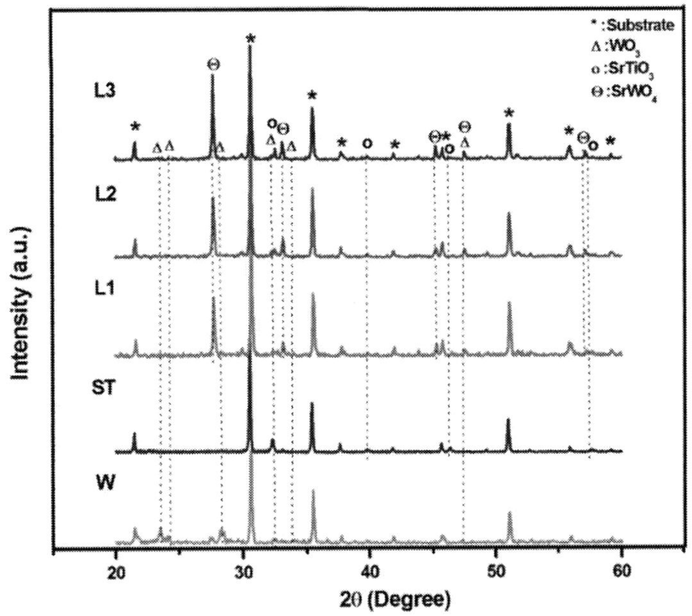

Fig. 1: X-ray diffraction pattern of various thin films

Fig. 2 presents the optical absorption spectra of various thin film samples. It was observed that the fundamental absorption edge of the $SrTiO_3/WO_3$ bilayered thin film samples shift from 365 nm (Absorption edge of $SrTiO_3$) and 421 nm (Absorption edge of WO_3) to 426 nm, resulting in enhanced absorption into visible region and red-shift was observed in all bilayered thin films. It may be attributed to the enhanced formation of mixed oxide ($SrWO_4$) at the interface in all bilayered $SrTiO_3/WO_3$ thin films.

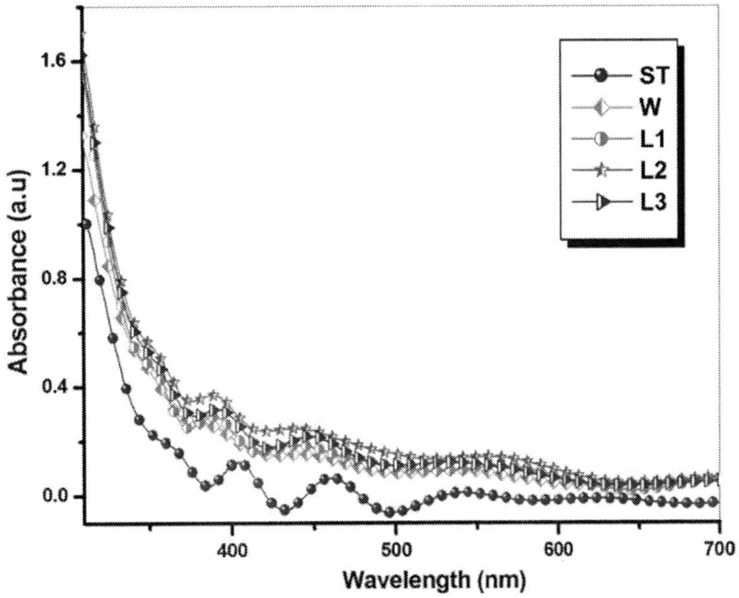

Fig. 2: Absorbance spectra for various thin films

Surface morphology of bilayered $SrTiO_3/WO_3$ thin film samples has been depicted in Fig.3. It was obseved that homogeneous WO_3 film was deposited on ITO substrate with good film-to-substrate adhesion. Rough surface with aggregation of grains is shown in overlying $SrTiO_3$ thin film. $SrTiO_3$ layer shows porous structure that may be beneficial for the diffusion of electrolyte and the effective scattering of incident light [8].

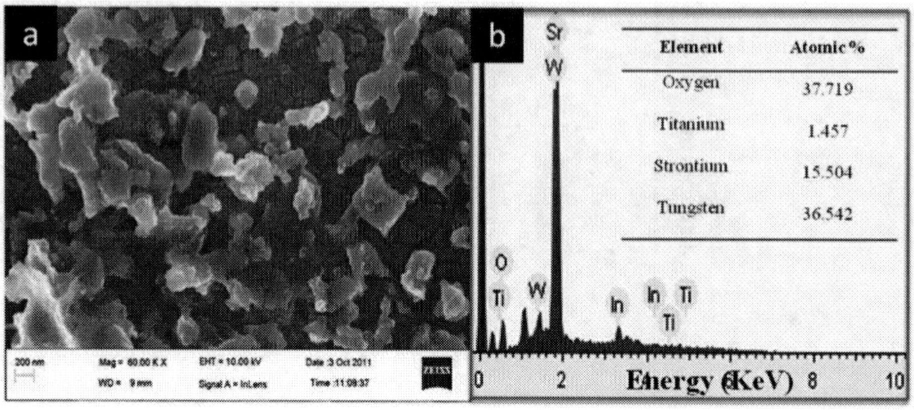

Fig. 3: Bilayered $SrTiO_3/WO_3$ thin film (a) SEM image (b) Energy dispersive X-ray map

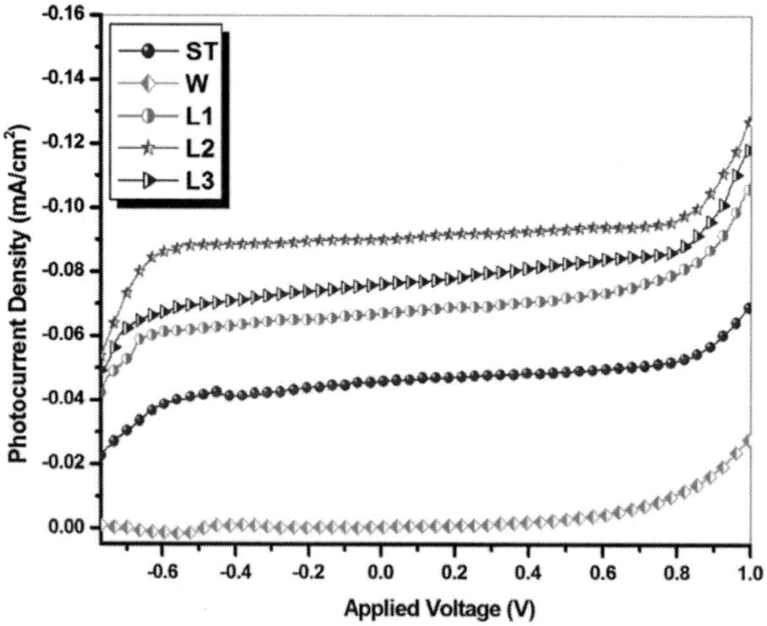

Fig. 4: Photocurrent density versus applied potential curve for various thin films

The photoelectrochemical application of various electrodes viz. $SrTiO_3$, WO_3 and $SrTiO_3/WO_3$ bilayered photoelectrode were investigated in a PEC cell, employing 150 W Xenon lamp (150 mW cm^{-2} illumination intensity) as light source for illumination. Photocurrent density of various samples was calculated by subtracting the current recorded under illumintion with that of current recorded under darkness, corresponding to same voltage and with same working electrode.

Fig. 4 shows photocurrent density versus applied voltage plot for $SrTiO_3$, WO_3 and all bilayered $SrTiO_3/WO_3$ thin film samples. It was observed that all bilayered $SrTiO_3/WO_3$ thin film samples shows higher photocurrent density than that of pristine $SrTiO_3$ and pristine WO_3. The photocurrent density increases with increase in thickness of $SrTiO_3$ (as seen in Sample L_2), but decreases after attaining a certain thickness (as seen in Sample L_3) [9].

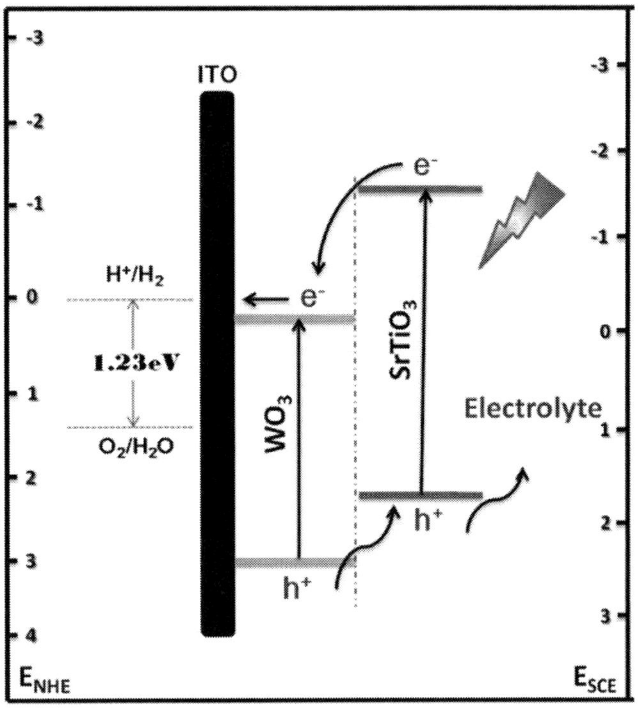

Fig. 5: Energy band diagram of Bilayered $SrTiO_3/WO_3$ thin film

These results can be further explained on the basis of charge carrier transfer across the interface [Fig. 5]. Upon illumination, electrons in both semiconductors get excited leaving holes in their corresponding valence band. The favourable conduction and valence band edge alignment lead to separation and transfer of electrons and holes to counter electrode and towards electrolyte respectively to perform water splitting reaction. Thereby both charge carriers separated effectively at the interface and generate an electrostatic field that may lead to improved photocurrent in all bilayered thin films even in the absence of external applied bias (i.e. at zero bias). However, the conduction band of underlying WO_3 lie below to the reduction potential of water, theoretically this condition is not suitable for hydrogen generation in unbiased condition but the significant formation of mixed oxide ($SrWO_4$) in all bilayered thin films (as confirmed by Fig.1) seems to be responsible for their improved photoelectrochemical response in the absence of an external bias.

Conclusion

Nanostructured n-n heterojunction were fabricated by depositing $SrTiO_3$ of varying thicknesses over WO_3 using sol-gel method. Enhanced photoelectrochemical activity of bilayered $SrTiO_3/WO_3$ thin films with respect to pristine $SrTiO_3$ and pristine WO_3 thin film was investigated. Bilayered $SrTiO_3/WO_3$ thin film (sample L_2) with 130 nm thick $SrTiO_3$ exhibited around two fold higher photocurrent density than that of pristine $SrTiO_3$ and much higher than that of pristine WO_3. Proper conduction and valence band edge alignment of both semiconductors, mixed oxide formation at the interface, improved open circuit voltage and inbuilt electric field at the interface seems to be the major contributing factors of improved photoresponse of bilayered $SrTiO_3/WO_3$ thin films in water splitting process.

Acknowledgements

The financial support received from University Grants Commission, New Delhi, India (Project No. 39-838/2010 (SR)) is acknowledged. The authors are thankful to Dr. Dinesh Deva of DST Unit of Nanosciences, IIT Kanpur for his help during SEM analysis.

References

1. Anghel M., Niculescu V-C., Stefanescu I., Tamaian R. (2010), "Biomass - an environmental friendly production source for hydrogen", *International Journal of Energy and Environment*, 4, 4.
2. Grätzel M. (2001), "Photoelectrochemical cells", *Nature*, 414, 338.
3. Currao A. (2007), "Photoelectrochemical Water Splitting", *Chimia*, 61, 815.
4. Hwang Y.J., Boukai A., Yang P. (2009), "High Density n-Si/n-TiO_2 Core/Shell Nanowire Arrays with Enhanced Photoactivity", *Nano Letters*, 9, 410.
5. Wang F., Valentin C.D., Pacchioni G. (2012), "Doping of WO_3 for Photocatalytic Water Splitting: Hints from Density Functional Theory", *J. Phys. Chem. C*, 116, 8901.
6. Zhu S., Zu X.T., Wang L.M., Ewing R.C. (2002), "Nanodomains of pyrochlore formed by Ti ion implantation in yttria-stabilized zirconia", *Applied Physics Letters*, 80, 23.
7. Shannon R.D. (1976),"Revised effective ionic radii and systematic studies of interatomic distances in halides and chalcogenides", *Acta Cryst,* A32, 751.
8. Dang T.C., Pham D.L., Nguyen H.L., Pham V.H. (2010), "CdS sensitized ZnO electrodes in photoelectrochemical cells", *Adv. Nat. Sci.: Nanosci. Nanotechnol.*, 1, 035010 (6pp).
9. Chi C-F, Lee Y-L, Weng H-S (2008), "A CdS-modified TiO_2 nanocrystalline photoanode for efficient hydrogen generation by visible light", *Nanotechnology*, 19, 125704 (5pp).

AIChE 2012 Session: Nanomaterials for Hydrogen Production and Fuel Cells

Title: Dopant Distribution and Activity in α-Fe_2O_3 for Water Splitting

Authors: C.D. Bohn, A. A. Talin, V.A. Szalai

Hematite is an attractive material for photoelectrochemical water splitting owing to its ~2 eV bandgap, stability in alkaline electrolytes and abundance. The interface between the semiconductor metal oxide and electrolyte is the location of primary interest for the overall water splitting process, yet methods for determining dopant concentration at this interface often give contradictory results. Here, techniques including energy dispersive X-ray spectroscopy (EDS), electron energy loss spectroscopy (EELS), electrochemical impedance spectroscopy, electronic conductivity and secondary ion mass spectroscopy are compared to determine dopant density. Discrepancies suggest that dopant activation, rather than simply increasing the dopant concentration, is key for achieving higher water splitting efficiencies. Specifically, secondary ion mass spectroscopy, which measures total Sn present, gives values of Sn concentration which are higher than that obtained from conductivity measurements, which measures electrically active Sn. Results are discussed in the broader context of photoelectrochemical hydrogen production and recommendations for improving the efficiency of water splitting with hematite are proposed.

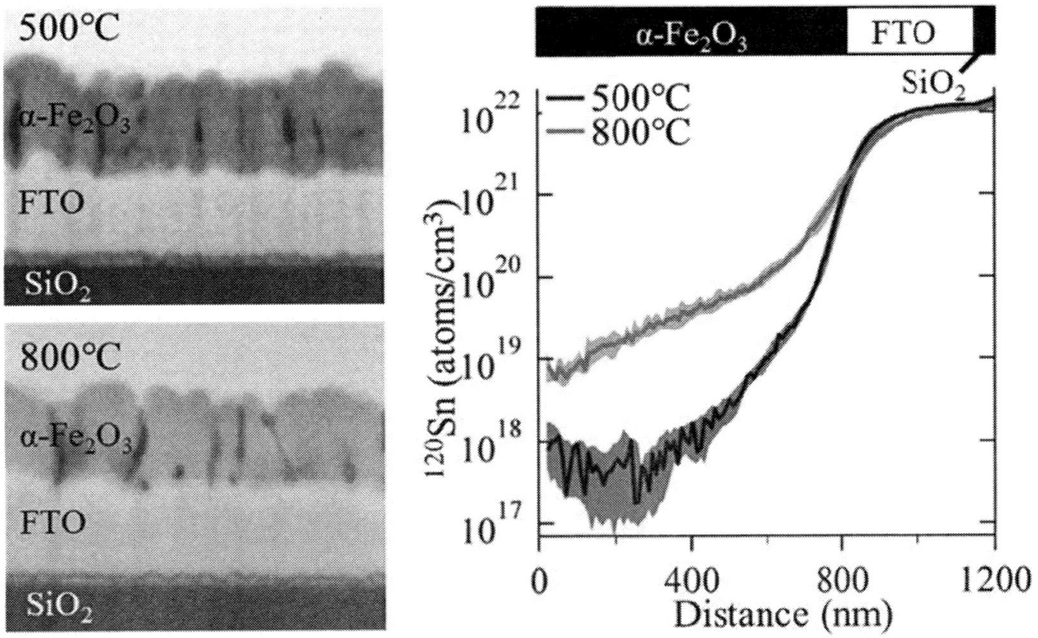

Figure 1. Scanning electron micrographs of cross-sectioned hematite sputtered films annealed at two different temperatures, along with dopant atom levels as a function of distance in the films.

HYDROGEN GENERATION USING COUPLED ZnO/TiO_2 PHOTOCATALYST OXIDES

A.menan M. Hussein[1], Luther Mahoney[2], Ranjit Koodali[2] and Rajesh Shende[1]
[1]South Dakota School of Mines and Technology, Rapid City, SD, USA
[2]University of South Dakota, Vermillion, SD, USA

Abstract

In this study, we designed and characterized suitable photo-active material of low-cost and improved efficiency in water splitting in the presence of a sacrificial agent. The performance of the coupled ZnO-TiO_2 photo-catalyst for hydrogen generation showed an enhanced activity when compared to a single oxide ZnO, and slightly improved performance when compared to TiO_2 photocatalysts. We hypothesized that the enhanced photo activity in water splitting was due to effective charge separation at the interface and delayed charge recombination rates, as a result of coupling and heterojunction interface. Coupled semiconductors with slightly staggered conduction band edges have an intrinsically built-in potential gradient at the interface which facilitates efficient charge separation and interfacial charge transfer. This potential gradient prevents premature photoelectron/hole recombination before a desired chemical reaction takes place on the surface of semiconductor particles. Ordered mesoporous coupled TiO_2/ZnO materials were synthesized using Pluronic P123 (P123) as the surfactant, titanium ethoxide and $ZnCl_2$ precursors. The materials obtained were characterized using XRD and their effectiveness towards hydrogen generation from aqueous methanol solution was measured under solar irradiation.

Introduction

With regard to hydrogen generation *via* water splitting using photo-catalysis, it could be alleviated to some extent with the help of implementing efficient photo-active materials. The photo-catalytic reactions are carried out in the presence of light and semiconductor. In the case of using the sunlight in its entirety, however, numerous challenges lie ahead in increasing hydrogen production efficiency. On the other hand, the charge separation and transport properties of single and/or conventional materials may not be sufficient for sustainable and economical energy harvesting.

A major drawback in achieving high photo-catalytic efficiency is the quick recombination of charge carriers. Recombination, which has faster kinetics than surface redox reactions, is a major drawback as it reduces the quantum efficiency of photo-catalysis [1, 2]. Therefore, it is an important way to promote the photo-catalysis efficiency by decreasing the recombination rate. Coupled and highly ordered mesoporous semiconductor materials offer a powerful approach for tailoring the transport property, charge separation ability, minimizing charge recombination and in this way a great enhancement in a photo-catalyst performance is expected. On the way towards improved photo-activity, new ways of preparing semiconductor composites are pursued and coupled ZnO-TiO_2 offers a new perspective in this veneration. In this study, we designed and characterized suitable photo-active material of low-cost and improved efficiency in water splitting in the presence of sacrificial agent. The performance of the coupled ZnO-TiO_2 photo-catalyst for hydrogen generation has shown an enhanced activity when compared to a single oxide ZnO and slightly improved comparing to pure TiO_2 photo-catalysts. We hypothesized that the enhanced photoactivity in water splitting due to effective charge separation at the interface as well as delayed charge recombination rates resulted from coupling and heterojunction interface. Advantages of coupled semiconductors with slightly staggered conduction band edges have intrinsically built-in potential gradient at the interface that facilitates efficient charge separation and interfacial charge transfer that prevents premature photoelectron/hole recombination before a desired chemical reaction takes place on the surface of semiconductor particles [3-5]. Furthermore, coupling of two semiconductors to create hetrojunctions, possessing different energy levels for their corresponding band and valence band, can also be used to enhance photocatalysis [6].

In most cases, coupled semiconductor oxide photo-catalysts, have activities superior to those of pure photo-catalyst oxides. A systematic study of the nature of interaction between component semiconductor oxides, mode of preparation, surface orientation and surface area, composition of the simple binary oxide systems in relation to its photo-catalytic activity in water splitting could give a clearer picture of the origin of modified photo-activity in them. A systematic study and effectiveness of different forms of coupled titania and zinc oxide compositions for water splitting has not been studied. The present research deals with such a systematic study on a coupled ZnO/TiO_2 system.

Experimental Methods

Synthesis of ZnO-TiO₂ nanocomposites

The $ZnO-TiO_2$ mixed oxides were synthesized by a sol-gel method. In a typical procedure, 3 ml of titanium(IV) n-butoxide previously dissolved in a solution containing 7 ml absolute ethanol, 0.3 ml concentrated hydrochloric acid and 15 wt% pluronic P123 surfactant. After stirring for 30 minutes, desired amount of zinc nitrate hexahydrate previously dissolved in 7 ml absolute ethanol was periodically added into the above solution. The obtained clear solution was continuously stirred for 30 minutes then 1 ml de-ionized water drop wise introduced and subsequently aged at room temperature for 12 hours forming a clear gel. It was then dried at 80°C. The resultant gel was calcined in air at 450 °C for 5 hours with the constant heating rate of 1 °C/min.

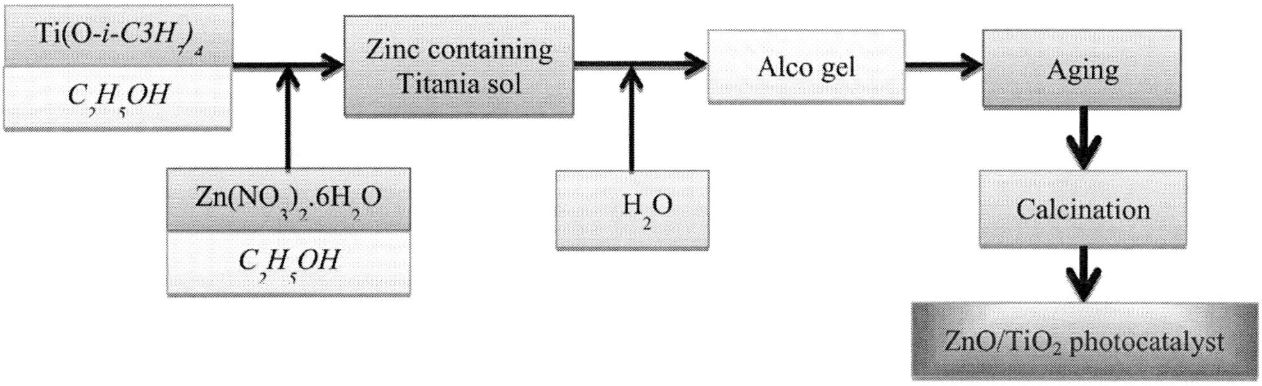

Figure 1: Schematic illustration of preparation method for coupled photocatalyst using conventional sol-gel method.

Photo-catalytic Testing Setup

A quartz reactor system, where water splitting reaction performed, is schematically depicted in Figure. 2. The set-up consisted of quartz reactor, filters and light source. The H_2 volume evolved from water splitting reaction was measured by manually injecting gas samples into SRI 8610C gas chromatograph (GC) provided with thermal conductivity detector (TCD) and molecular sieve column.

Figure 21: Schematic of photocatalytic reactor set-up.

Results and Discussion

XRD analysis of ZnO-TiO₂ composite

Coupled ZnO-TiO_2 photocatalysts containing different wt % Zn^{2+} loadings were prepared by the sol-gel method. The powder X-ray diffraction (XRD) patterns of the ZnO-TiO_2 composites containing different amounts of Zn^{2+} are shown in Figure 3, for 2θ diffraction angles between 10° and 80°. For comparison, the XRD pattern of nominally phase pure TiO_2 and ZnO are also included in Figure 3. The 2θ positions corresponding to 25°, 37° and 48° were attributed to different diffraction planes of anatase TiO_2 whereas 2θ positions of 33.5°, 35° and 36.5° matched with reflections of ZnO as also indicated by the ICDD patterns.

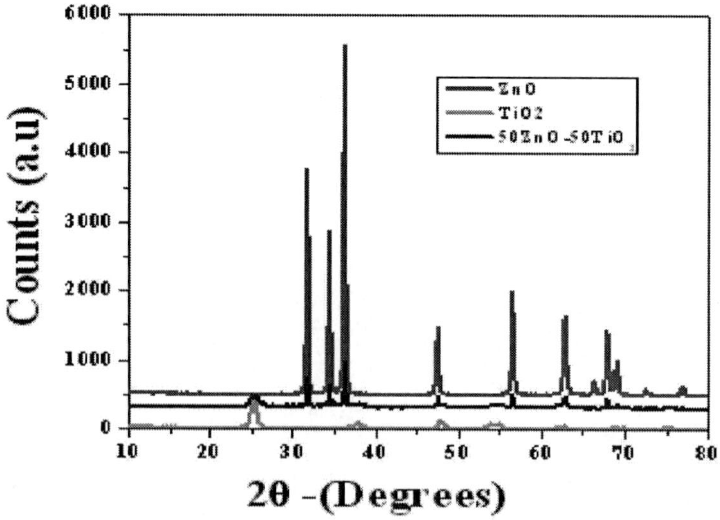

Figure 3: X-ray diffraction of anatase TiO_2, ZnO and ZnO-TiO_2 composites.

Photocatalytic Activity of ZnO-TiO₂ Composite

Figure 4 shows the photocatalytic activity of ZnO-TiO_2 nanocomposites containing Zn^{2+} loading in different amounts on the rate of hydrogen generation from aqueous CH_3OH solution

(20 vol%). It can be observed that the photocatalytic hydrogen production rate varies with the ratio of ZnO in ZnO/TiO_2 composite. The rates of hydrogen evolution increase along with Zn^{2+} content from 10 to 50 wt%; reaches a maximum at 50 wt% ZnO loadings. In contrast, further increase in Zn loading resulted in sharp decrease in photocatalytic hydrogen generation. In addition, nominally phase pure ZnO alone showed no measureable hydrogen evolution.

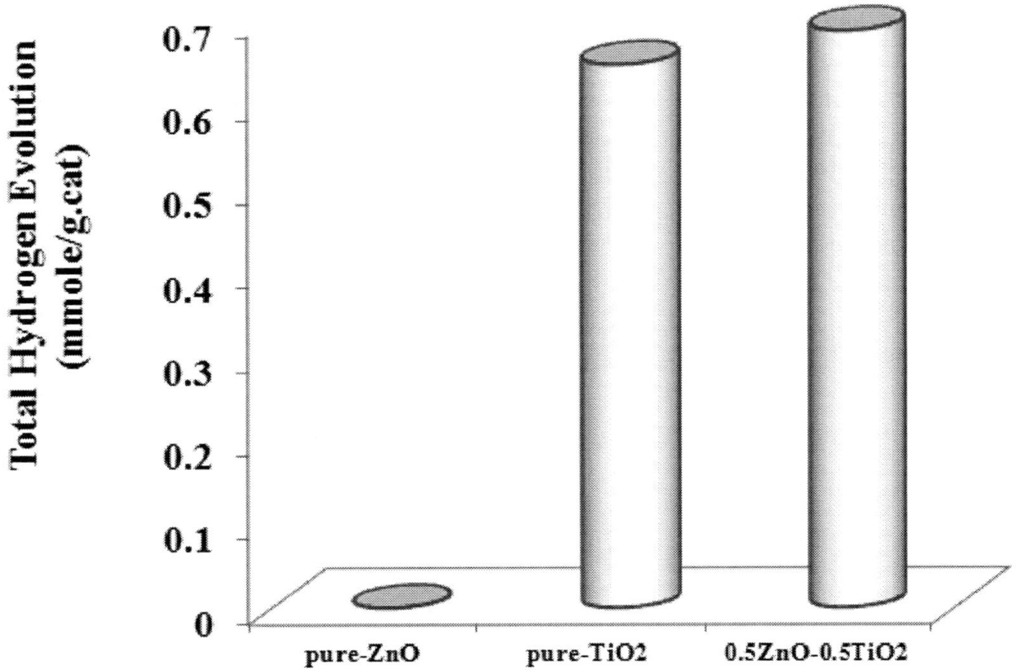

Figure 4: Photocatalytic activity measured for $ZnO-TiO_2$ composites at different Zn^{2+} loadings.

The photocatalytic materials prepared here were analyzed using impedance analyzer, which indicated appreciable difference in the charge transfer resistance in a bulk TiO_2 with and without ZnO composite (data not included). The charge transfer resistance to the electron transfer decreased appreciably with the introduction of ZnO in comparison with TiO_2 alone. Based on this measurement, we believe in the following mechanism responsible for charge separation and photoactivity of $ZnO-TiO_2$ composite (Figure 5). Under UV light irradiation, both TiO_2 and ZnO are highly efficient photocatalysts, however, greatly restricted due the bottlenecks that reduce photocatalytic efficiencies. Light incident on the surface of $ZnO-TiO_2$ induces a charge separation between excited electron and its valance band hole, where electrons transfer from the ZnO conduction band to the TiO_2 conduction band and conversely, holes transfer from the TiO_2 valence band to the ZnO valence band. The presence of a $ZnO-TiO_2$

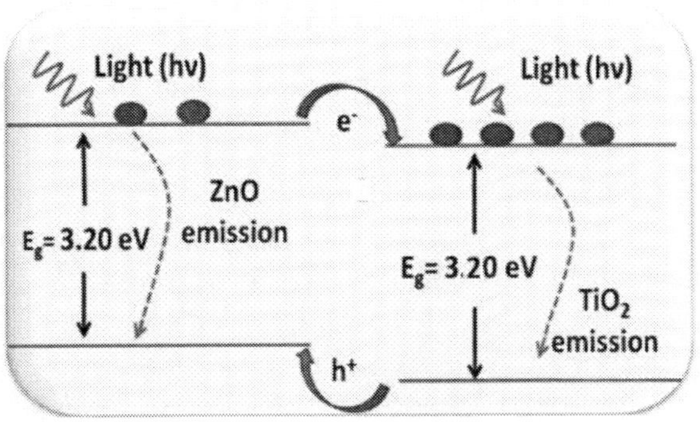

Figure 5: A schematic diagram illustrating the principle of electron separation and photoactivity at $ZnO-TiO_2$ hetero-junction system.

heterojunction as the potential barrier limits the chance for the recombination of electron–hole pairs. This increases the availability of the electrons or holes to migrate to the TiO_2 or ZnO surface of the ZnO-TiO_2 composite. Consequently improves the occurrence of redox processes. Such an efficient charge separation increases the lifetime of the charge carriers and increases the efficiency of the interfacial charge transfer to adsorbed substrates.

Conclusions

The enhancement in activity has been attributed to various causes, in most cases: nature of added oxide, coordination, mode of preparation and treatment, surface area, composition, and nature of interaction the component oxides. 50 ZnO-50 TiO_2 composite produces more hydrogen with respect TiO_2 alone. This photo-catalytic enhancement could be attributed to the formation of an active interface between ZnO-TiO_2 hetero-junction, high surface area and large pore volume. The bandgap calculated from diffusive reflectance spectra shows that coupled ZnO-TiO_2 has higher bandgap with respect to bare ZnO or TiO_2. Therefore, it is clear that while a suitable band-gap is essential for photo-catalytic efficiency, it is not the sole criterion.

Acknowledgements

The authors acknowledge the financial and research assistantship support of NSF–SD IGERT, DGE-0903685.

References

1. Youngblood, W. J.; Lee, S.-H. A.; Maeda, K.; Mallouk, T. E., Visible Light Water Splitting Using Dye-Sensitized Oxide Semiconductors. *Accounts of Chemical Research* 2009, *42* (12), 1966-1973.
2. Maolin Zhang, Taicheng An, Xiaolu Liu, Xiaohong Hu, Guoying Sheng, Jiamo Fu, Preparation of a high-activity ZnO/TiO2 photocatalyst via homogeneous hydrolysis method with low temperature crystallization, Materials Letters 64, 2010. 1883–1886.
3. Akira Fujishima, Tata N. Rao, Donald A. Try, Titanium dioxide photocatalysis, Journal of Photochemistry and Photobiology C: Photochemistry Reviews 1, 2000, 1–21
4. D. Schlettwein, T. Oekermann, T. Yoshida, M. Tochimoto, H. Minoura, Photoelectrochemical sensitization ZnO–tetrasulfophthalocyaninatozinc composites prepared by electrochemical self-assembly, Journal of Electroanalytical Chemistry 481, 2000, 42–51.
5. Yanfa Yan, K.-S. Ahn, T. Deutsch, M. Huda, S. Shet, S.H. Wei, J. Turner, and M.M. Al-Jassim, Application of ZnO for photoelectrochemical splitting of water. 213th ECS Meeting, Abstract #780, © The Electrochemical Society.
6. Wan-Kuen Jo and Kun-Hwan Kim, Feasibility of Carbonaceous Nanomaterial-Assisted Photocatalysts Calcined at Different Temperatures for Indoor Air Applications, International Journal of Photoenergy, 2012, 2012, pp. 9

(511f) Solution Combustion Synthesis of Nanostructured Cobalt Oxide for Sodium Borohydride Hydrolysis

Lori J. Groven[1], Teandra L. Pfeil[1] and Timothee L. Pourpoint[2], (1)School of Mechanical Engineering, Purdue University, West Lafayette, IN, (2)School of Aeronautics and Astronautics, Purdue University, West Lafayette, IN

Cobalt oxide and $LiCoO_2$ have recently been explored as catalysts for the hydrolysis of sodium borohydride and ammonia borane with promising results. In several reports the hydrogen generation rate may meet that of typical, more expensive, noble metal catalysts such as Ru, Rh and Pt. However, the synthesis of such materials has generally been from decomposition of nitrates, acetates and mixtures thereof and can be quite slow (hours). This work reports the solution combustion synthesis of nanostructured cobalt oxide catalyst powder from cobalt nitrate-glycine solutions. The performance characterization of the catalyst for hydrolysis of sodium borohydride is compared to commercial Co and Co_3O_4 nanopowders. Solution combustion synthesis is a cost effective, fast (minutes), reproducible method to synthesize powders with high purity (98-99%) and surface areas comparable to commercially available Co_3O_4 nanopowders (26-32 m^2/g). While X-ray diffraction analysis showed that the synthesized powders were not highly crystalline, the hydrogen generation rate, at 5 wt.% catalyst loading, was double that of commercial Co_3O_4 and Co nanopowders. Catalyst durability/reutilization will be discussed in detail.

(576a) Noble Metal-Based 1-Dimensional Nanomaterials in the Application of Direct Alcohol Fuel Cells

Liang Su, Yixin Liu, Wenzhao Jia and Yu Lei, Chemical, Materials & Biomolecular Engineering, University of Connecticut, Storrs, CT

Direct alcohol fuel cells (DAFCs) are envisaged to be a new generation of portable power sources which convert chemical energy into electrical energy with, theoretically, both economical and environmental benefit. Noble metal-based composite nanomaterials have shown great applicability as the cathode and anode catalysts in application of DAFCs. Furthermore, 1-D bimetallic nanomaterials have shown improved performance towards oxygen reduction reaction and alcohol oxidation reaction attributed to the ameliorated durability from the 1-D nanostructure and the enhanced catalytic activity from the incorporation of the second metal.

In this work, Pt-based nanomaterials with 1-D hollow nanostructure prepared from a novel, robust and versatile electrospinning–electroless plating–galvanic replacement reaction process. The overall procedure for the material preparation is also economically efficient and environmentally friendly. The as-prepared Pt nanotubes and Pt-carbon nanofibers have been characterized by SEM, TEM, XRD, and XPS. Furthermore, these nanomaterials have shown excellent catalytic activity with superior durability towards oxygen reduction reaction and ethanol oxidation reaction in the application of DAFCs.

(576b) Nanoporous, Conducting, Electrocatalytically Active, 10nm Thick Core-Shell Au@Pt Nanoparticle Bilayer Architectures for PEM Fuel Cell

Ipshita Banerjee, Chemical Engineering, Indian Institute of Science, Bangalore, India, Venugopal Santhanam, Department of Chemical Engineering, Indian Institute of Science, Bangalore, India and V Kumaran, Chemical Engineering, Indian Institute of science, Bangalore, India

The main challenge facing the development of Proton exchange membrane fuel cells (PEMFC) currently is to fabricate a catalyst layer with minimum loading of precious platinum nanoparticles, maximum availability of eletrocatalytically active area, and high durability of the catalyst layer in presence of poisonous CO species and corrosive electrochemical atmosphere. The different approaches reported in literature to prepare catalyst layers can be broadly divided into two types: 1) where the platinum nanoparticles are physisorbed on carbon substrates like carbon nanotubes, buckypaper, xerogels etc., and 2) where thin films of white gold leaf are dealloyed and electroplated with a thin platinum layer to form the electrocatalytically active catalyst membrane. In our group, we are developing a novel bottom-up approach, wherein self-assembled monolayers of alkanethiol-coated, monodispersed Au@Pt core shell nanoparticles are stacked to fabricate the catalyst layer. Herein, results on bilayer membranes of Au@Pt core shell nanoparticles will be reported. Initially, the bilayers were electrocatalytically inactive, and had a high sheet resistance of the order of GΩ, due to the presence of organic ligands. The bilayers were then treated to room temperature RF plasma in an argon environment. The results of electrical, spectroscopic and microscopic characterization show that argon plasma exposure removes the alkanethiol ligand coating and leads to coalescence of neighbouring nanoparticles, thereby rendering the bilayer electrocatalytically active and electronically conducting. After an optimized time of exposure, the individual nanoparticles in the bilayer stacked laterally coalesce together to form electrically continuous, nanoporous metallic film with sheet resistances that are eight orders of magnitude lower than the ligand-coated bilayer array. We will present the results of our studies on the electrocatalytic activity and conductivity of membranes prepared with different Pt/Au ratios, and for different durations of argon plasma exposure. The activity of these bilayer membranes towards methanol oxidation and CO oxidation will also be reported.

(576c) Ion Pair Reinforced Semi-Interpenetrating Polymer Networks for Fuel Cell Polymer Electrolyte Membrane Design

Chunliu Fang, Graduate School for Integrative Science and Engineering, National University of Singapore, Singapore, Singapore, Liang Hong, Department of Chemical and Biomolecular Engineering, National University of Singapore, Singapore, Singapore and **Jim Yang Lee**, Chemical & Biomolecular Engineering, National University of Singapore, Singapore, Singapore

Semi-interpenetrating polymer networks (SIPN) are a special class of polymer composites where one or more linear (or branched) polymers percolate molecularly into a network of other polymers. It provides a high degree of homogenization of the component polymers beneficial for the fabrication of polymer electrolyte membranes (PEMs) for fuel cell applications. For example a linear ionomer can be used as the proton source to infiltrate into a cross-linked hydrophobic polymer network which provides the mechanical properties. The interlocked polymer structure leads to good intermixing of the hydrophilic and hydrophobic components to support a distributed presence and good connectedness of the ionic clusters.

Despite these promising prospects, SIPN membranes for fuel cell applications have yet to be satisfactorily demonstrated. In many of the studies, the network polymer was cross-linked by ester linkages. The susceptibility of the ester linkages to hydrolytic degradation in the acidic fuel cell environment compromises the SIPN performance.

This study proposes a new SIPN design where ion pairs are used to reinforce the SIPN structure. The SIPN was synthesized from linear sulfonated poly(2,6-dimethyl-1,4-phenylene oxide) (SPPO), brominated PPO (BPPO), and ethylenediamine (EDA) cross-linker by a one-step thermal cross-linking procedure. Ion pairs were formed after cross-linking and strengthened the attachment of SPPO to the BPPO/EDA network. The cross-links in the membranes are chemically resistant. Measurements of membrane dimensional changes and hydrolytic stability confirmed the improvements made to the PPO membranes for fuel cell applications.

(576e) Nanoscale Membrane Design Principles for Optimizing Proton Conductivity

David J. Keffer, Chemical and Biomolecular Engineering, University of Tennessee, Knoxville, Knoxville, TN

A model for charge transport in aqueous systems is used to provide insights into design of proton exchange membranes that optimize charge transport. The model accounts for the effect of acidity, confinement and connectivity on the diffusivity and has been shown to quantitatively agree with experimental measurements of charge self-diffusivity in Nafion [Esai Selvan, M., et al. *J. Phys. Chem. B* **115** 2011 pp 3052–3061] and in cross-linked and sulfonated polycyclohexadiene. In this presentation, the origin of the theoretical model is described. The model is applied to water-filled cylindrical nanopores functionalized on their interior surface with acid groups. It is demonstrated that for cylindrical nanopores of a given radius there is an optimal surface coverage of acid groups. The optimum can be sharply peaked, indicating that non-optimal surface coverages (either too low or too high) drastically reduce the conductivity of the pore. The theoretical maximum conductivity through a cylindrical nanopore is calculated to be about 0.70 S/cm at 300 K. Comparison with available experimental results is made. Limitations of the model are also discussed.

(751a) Simulation of Lithium-Ion Cells and Batteries

Ralph E. White, Chemical Engineering, University of South Carolina, Columbia, SC

Simulation of lithium-ion cells and batteries includes predicting the performance and life of these devices. The beginning of life performance of lithium-ion cells can be predicted by using John Newman's porous electrode theory, which will be reviewed in this presentation. Predicting the rate of heat generation in lithium-ion cells is a critical aspect of designing heat removal components of a lithium-ion battery pack for electrical vehicle applications. The rate of heat generation of the components in a lithium ion cell can be predicted by using multiscale, multidimensional modeling technique, which will be presented in this talk. In order the reduce the computation burden for solving these model equations, reduced order models have been developed and will be outlined in this tutorial. The life of lithium-ion cells and batteries can be predicted by adding equations that describe the degradation processes to the model equations, or by using parameter estimation techniques to determine how the important physical parameters in the beginning of life model equations change with time. An overview of these life predicting techniques will be presented.

(751b) Design Aspects of Flow Batteries for Large Scale Energy Storage

Robert Savinell, Chemical Engineering, Case Western Reserve University, Cleveland, OH

Large scale energy storage technology is needed to facilitate integrating renewable energy generation technologies, such as wind and solar, into the electric grid because these energy generation technologies are inherently intermittent and thus matching supply to load demand is a challenge. Also, energy storage is needed for modernizing the grid into a "smart" electricity distribution system. There is a large variety of potential energy storage solutions available for this broad range of applications and understanding the various operating and economic constraints and requirements is necessary to choose a technology. Flow batteries of various chemistries are gaining attention again in recent years because of these markets and opportunities and of the inherent advantage of flow batteries. The advantages and disadvantages, and their technical and economic challenges will be addressed for a number of approaches. The presentation will focus its attention of electrode materials, electrode design, and cell design for redox flow batteries.

(751c) Electrode and Cell Design Principles in Batteries and Fuel Cells

Trung V. Nguyen, University of Kansas, Lawrence, KS

In a battery or fuel cell, the electrodes and other components in the cell must be carefully selected and designed to address the electrode potentials, reactions, and transport (electronic, ionic, molecular) processes, especially the limiting ones, of the system to meet the performance (discharge and charge rate and cycle life), shelf-life or storage life, and safety requirements. While high active surface area is needed for high power, its effect on the self-discharge rate and corrosion should also be considered. In addition to the active materials that participate in the electrochemical reactions, the role of the supporting materials such as the current collectors, catalyst supports, electrolytes, and binders, both conductive and non-conductive, must also be considered because they can have a significant effect on the performance and/or shelf-life of the system. This presentation will discuss the materials selection and components and cell design requirements for a few battery and fuel cell systems.

(751d) Electrowinning Principles and Practice: A Brief Introduction to the Electrochemistry and Engineering Behind Modern Electrowinning Plants

Dan Steingart, Energy Institute and Chemical Engineering, City College of New York, New York, NY

Electrowinning, or electroextraction, is the oldest industrially applied electrochemical process, and the backbone of the production of many metals, both precious and otherwise (aluminum, copper, zinc and lead, among others), as well as industrial gases (e.g. the chlor akali process). In the US alone electrowinning processes consume on the order of 1 to 2% of total electricity produced, and many of the most energy intensive processes are have much room for improvement of efficiency. In this tutorial we will broadly examine the theory and practice of both the electrochemical processes as well as the supporting mining and extraction operations. Beyond a discussion of the basic operating principles, we will examine issues of feed purity, flow control, and rate sensitivity. Finally, experimental and R&D scale approaches to efficiency improvements will be covered. While the aluminum, copper, and chlor-alkali systems will be systems of focus in this talk, the principles discussed will be applicable to any modern electrowinning process.

(751e) Design of Enzyme Electrodes for Sensing and Power Applications

Joshua Gallaway, Chemical Engineering, Energy Institute, City College of New York, New York, NY

The prospect of harnessing the catalytic functionality of enzymes for practical applications holds tremendous scientific and technological promise. Composed of folded proteins built around an active site, enzymes are biological catalysts that enable thousands of highly specific reactions to occur under mild conditions. In biological systems, electron exchange to and from enzymes happens diffusionally from small redox molecules. In contrast, practical systems designed by humankind often rely on electron exchange at a heterogeneous interface (an electrode). By using mediator molecules to transport electrons to enzymes they can be used on electrodes with relatively high currents, but the resulting systems can be complex. Electron hopping, Michaelis-Menton kinetics, materials aspects of hydrogels, and other complications make enzyme electrode design less than intuitive. This talk aims to demonstrate the connection between electrode modeling, synthesis techniques, and fabrication, which when used together can greatly improve the practical performance of enzyme electrodes. An example case will be presented in which current response of an oxygen reducing enzyme electrode was doubled by molecular engineering.

AUTHOR INDEX

AUTHOR INDEX